Mechanical Desktop 2.0:
Applying Designer and Assembly Modules

Mechanical Desktop 2.0:
Applying Designer and Assembly Modules

Daniel T. Banach

Press

an International Thomson Publishing company I(T)P®

Albany • Bonn • Boston • Cincinnati • Detroit • London • Madrid
Melbourne • Mexico City • New York • Pacific Grove • Paris • San Francisco
Singapore • Tokyo • Toronto • Washington

NOTICE TO THE READER

Publisher does not warrant or guarantee any of the products described herein or perform any independent analysis in connection with any of the product information contained herein. Publisher does not assume, and expressly disclaims, any obligation to obtain and include information other than that provided to it by the manufacturer.

The reader is expressly warned to consider and adopt all safety precautions that might be indicated by the activities herein and to avoid all potential hazards. By following the instructions contained herein, the reader willingly assumes all risks in connection with such instructions.

The publisher makes no representation or warranties of any kind, including but not limited to, the warranties of fitness for particular purpose or merchantability, nor are any such representations implied with respect to the material set forth herein, and the publisher takes no responsibility with respect to such material. The publisher shall not be liable for any special, consequential, or exemplary damages resulting, in whole or part, from the readers' use of, or reliance upon, this material. Autodesk does not guarantee the performance of the software and Autodesk assumes no responsibility or liability for the performance of the software or for errors in this manual.

Trademarks
Mechanical Desktop and the Mechanical Desktop® logo are registered trademarks of Autodesk, Inc. AutoCAD and the AutoCAD® logo are registered trademarks of Autodesk, Inc. Windows is a trademark of the Microsoft Corporation. All other product names are acknowledged as trademarks of their respective owners.

Autodesk Press Staff
Publisher: Michael A. McDermott
Acquisitions Editor: Sandy Clark
Production Coordinator: Jennifer Gaines
Art and Design Coordinator: Mary Beth Vought
Editorial Assistant: Christopher Leonard

Cover: Image© 1998 Autodesk, Inc.

For more information, contact

Autodesk Press
3 Columbia Circle, Box 15-015
Albany, New York   USA 12212-5015

International Thomson Publishing Europe
Berkshire House 168-173
High Holborn
London, WC1V 7AA
United Kingdom

Thomas Nelson Australia
102 Dodds Street
South Melbourne, Victoria 3205
Australia

Nelson Canada
1120 Birchmont Road
Scarborough, Ontario
Canada, M1K 5G4

International Thomson Publishing Southern Africa
Building 18, Constantia Park
240 Old Pretoria Road
P.O. Box 2459
Halfway House, 1685 South Africa

COPYRIGHT© 1998 Delmar Publishers
Autodesk Press imprint
an International Thomson Publishing Company
The ITP logo is a trademark under license.
Printed in the United States of America

International Thomson Editores
Campos Eliseos 385, Piso 7
Colonia Polanco
11560 Mexico D. F.   Mexico

International Thomson Publishing GmbH
Konigswinterer Strasse 418
53227 Bonn   Germany

International Thomson Publishing France
Tour Maine-Montparnasse
33, Avenue du Maine
75755 Paris Cedex 15, France

International Thomson Publishing–Japan
Hirakawacho Kyowa Building, 3F
2-2-1 Hirakawa-cho Chiyoda-ku
Tokyo 102 Japan

International Thomson Publishing Asia
60 Albert Street
15-01 Albert Complex
Singapore 189969

All rights reserved. No part of this work covered by the copyright hereon may be reproduced or used in any form or by any means—graphic, electronic, or mechanical, including photocopying, recording, taping, or information storage and retrieval systems—without the written permission of the publisher.

1 2 3 4 5 6 7 8 9 10 XXX 04 03 02 01 00 99 98

Library of Congress Cataloging-in-Publication Data
Banach, Daniel T.
   Mechanical desktop 2.0 : applying designer and assembly modules /
   Daniel T. Banach.
      p.   cm.
   Includes index.
   ISBN 0-7668-0068-7
   1. Engineering graphics. 2. Autodesk Mechanical desktop.
3. Engineering design—Data processing. 4. AutoCAD (Computer file)
I. Title
T353.B37 1998
620'.0042'02855369—dc21   98-14979
                                                                                                                   CIP

contents

Acknowledgments ix

Dedication xi

Introduction xiii

Chapter 1—Sketching, Profiling, Constraining, and Dimensioning 1

Specifying the Part Settings 1

Sketch the Outline of the Part: Step 1..3

Profile the Sketch: Step 2...5

Constrain the Sketch: Step 3 7

Dimension the Sketch Profile: Step 4 13

Review Questions...22

Chapter 2—Viewing, Extruding, Revolving, Sweeping, and Editing Parts........23

Viewing a Model from Different Viewpoints...23

Using the XY Orientation in 3D 26

Extruding the Profile 27

Revolving the Profile..32

Sweeping the Profile 38

Editing Features in Parts (Feature Editing) 45

Review Questions...55

Chapter 3—Work Axis, Sketch Planes, Work Planes, Work Points and Visibility Options 57

Creating a Work Axis 57

Making a Plane the Active Sketch Plane 59

Creating a Work Plane 65

Creating Work Points 78

Controlling the Visibility of Objects 79

Review Questions 82

Chapter 4—Cut, Join and Intersect Operations, Browser Actions, Fillets, Chamfers, Holes and Arrays 83

What Is a Feature? 83

Cut, Join and Intersect Operations 84

Using the Browser for Creating and Editing 101

Fillets 108

Chamfers 117

Holes 123

Arrays 135

Review Questions 145

Chapter 5—Advanced Dimensioning, Constraining and Sketching Techniques 147

Dimension Display and Equations 147

Design Variables 153

Close Edge 163

Copy Sketch 166

Converting Existing 2D Drawings to 3D Parametric Parts 172

Review Questions 177

Chapter 6—Advanced Modeling Techniques 179

Adding a Part to a Drawing 179

Mirroring a Part 183

Scaling a Part 186

Shelling 188

Copying a Feature 194

Reordering 197

Creating Combined Parts 200

Replaying 204

Mass Properties Information for the Active Part 207

Mass Properties for Multiple Parts 208

Review Questions 211

Chapter 7—Assemblies 213

Creating Assemblies 213

Top Down Approach 214

Bottom Up Approach 218

Assembly Constraints 221

Editing Assembly Constraints 242

Interference Checking 245

Creating Scenes 247

Tweaking Parts in a Scene 249

Creating Trails 250

Review Questions..256

Chapter 8—Drawing Views and Annotations..257

Creating Drawing Views 257

Editing Drawing Views 268

Moving Drawing Views 269

Deleting Drawing Views 270

Editing Dimensions..273

Creating a Bill of Materials..288

Review Questions..296

Chapter 9—Practice Exercises 297

Chapter 10—Introduction to NURBS Surfaces 337

Review Questions..357

Glossary 358

Index 369

acknowledgments

The author would like to thank the team at Autodesk Press for their efforts and encouragement, specifically Sandy Clark, Christopher Leonard, and Mary Beth Vought. Thanks also goes to the Autodesk MCAD group in Novi and Chicago for their assistance in getting information.

The author would also like to thank his family for their patience and understanding.

The author and publisher would like to thank and acknowledge the many professionals who reviewed the manuscript. A special acknowledgement is due to the following instructors and professionals, who reviewed the chapters in detail:

Jeffery R. Gibbs
Muskingum Area Technical College, Zanesville, OH

Robert Jones
University of Regina, Regina, SK

David Pitzer
Sonoma, CA

Gary J. Poulsen
Salt Lake Community College, Salt Lake City, UT

Mike Wheaton
ITT-Portland, Portland, OR

A special thank you to Bob Henry of Autodesk, Inc., Novi, MI for his careful and thoughtful technical editing.

about the author

Daniel T. Banach is the Application Engineering Manager at MasterGraphics, Wisconsin's leading Autodesk dealer, where he has taught AutoCAD and Mechanical Desktop classes. Since 1993, he has participated in the Autodesk Mechanical Desktop Gunslinger program, and many beta-testing programs. Dan is a renowned lecturer who has appeared at numerous seminars, including at Autodesk University and Design World. Dan also has five years of experience as a mechanical designer and drafter, where he started using AutoCAD in 1988, and holds a BS degree from UW Stout in Industrial Technology and an AS degree from Milwaukee Area Technical College in Photography.

dedication

To my wife, for without her love and support this book would not have been possible; and to my parents, who have always stood behind me.

introduction

Mechanical Desktop 2.0: Applying Designer and Assembly Modules

Welcome! If you are new to Mechanical Desktop or 3D design, you have just joined over 100,000 people already doing 3D design. If you are a current Mechanical Desktop user, you will find major enhancements in the software over the previous release. Look for an asterisk in front of the new and enhanced commands in the table of contents.

The chapters in this book follow the order in which you will create your own models and drawings. Each chapter introduces a set of topics and then takes you through a basic, step-by-step example. Each chapter builds on the material learned in the previous chapter(s). At the end of most chapters you will find practice exercises for you to complete on your own. They are based on real world parts used in different disciplines of design.

Product Background

Mechanical Desktop 2.0 was written by Autodesk and runs inside AutoCAD R14. Mechanical Desktop is a 3D feature-based parametric solid modeler that allows you to create complex 3D parametric models and to generate 2D views from those models.

Mechanical Desktop consists of AutoCAD R14 and four modules:

Designer: Feature-based parametric modeler. Part Modeling.

AutoSurf: Non Uniform Rational B-Splines (NURBS) surfaces. Surface Modeling.

Assembly: Manage and constrain assembled parts. Assembly Modeling.

Drawing Manager: 2D view layout and dimensioning for outputting engineering drawings.

All the functionality of AutoCAD Mechanical is found in Mechanical Desktop 2.0. This includes breaking, joining, aligning and inserting dimensions. Weld symbols, surface texture symbols and GDT symbol symbols are included.

Requirements

This book assumes that you are running Mechanical Desktop 2.0 and that you are proficient with AutoCAD commands such as lines, arcs, circles, polylines, move, erase, grips etc. If you are not proficient in those areas, you may want to refer to the AutoCAD online help as needed.

New Features

Since Mechanical Desktop runs inside AutoCAD, you do not need to relearn an entire new CAD package. Instead, you will build on your AutoCAD knowledge by learning around ninety new commands. This is the second release of Mechanical Desktop. Major new features are listed below.

Parametric Booleans: Two Designer parts or solids can now be cut, joined or intersected together, maintaining parametrics of both parts.

New interface: Mechanical Desktop commands have been streamlined and there is now the Desktop browser, which allows quick editing of parts and access to almost 85 percent of the Mechanical Desktop functionality.

Shelling: Ability to create thin-walled parts from a solid model.

Complex blending: Linear and cubic radius fillets as well as "n-sided" patches for blending multiple edges together.

Feature reordering: Features can be repositioned in the database hierarchy with a drag and drop technique through the browser.

Assembly methodology: Assemblies now use intelligent constraints, which allow you to cycle through the possible selection set on the screen until the correct option is highlighted. Mechanical Desktop constraint solver is now variational, which allows linkages to be assembled.

Expanded drafting tools: Mechanical Desktop now supports broken views, along with all the capabilities of AutoCAD Mechanical, which help you create welding, surface and G.D.T. symbols. The ability to break extension lines and maintain associativity is included.

Table-driven parts: Excel spreadsheets can now be linked to a Mechanical Desktop part to drive the parametric part, also known as charted or tabulated drawings.

Change from Mechanical Desktop 1.X

A major change has occurred in creating assemblies—you no longer have to make parts into components. When created, each part is automatically made into a component. If you are a current user and you have used components in your assemblies, your files will automatically be updated to Mechanical Desktop 2.0 files. If you used regular Designer parts in assemblies that have not been componentized, a wizard will appear when you open the file and guide you through your options in converting them to Mechanical Desktop 2.0 parts.

Basics of 3D modeling

If you are new to creating 3D models, you need to take time to evaluate what you are going to model and how you are going to approach it. When I evaluate a model, I look for the main basic shape. Is it flat or cylindrical in shape? Depending on the shape, I will take a different approach to the model. I try to start with a flat face if possible; from a flat face it is easier to add other features. If the model is cylindrical in shape, I look for the main profile or shape of the part and revolve or extrude that profile. After the main body is created, work on the other features, looking for how this shape will connect to the first part. Think about 3D modeling like working with building blocks, each block sitting on another block, but remember that material can also be removed from the original solid.

Terms and Phrases

To help you to better understand Mechanical Desktop, a few of the terms and phrases that will be used in the book are explained below.

Parametric Modeling: Parametric modeling is the ability to drive the size of the geometry by dimensions. For example, if you want to increase the length of a plate from 5" to 6", change the 5" dimension to 6" and the geometry will update. Think of it as the geometry along for a ride, driven by the dimensions. This is opposite to AutoCAD 2D dimensioning, known as associative dimensioning: as lines, arcs and circles are drawn, they are created to the exact length or size; when they are dimensioned, the dimension reflects the exact value of the geometry. If you want to change the size of the geometry, you stretch the geometry and the dimension automatically gets updated. Think of this as the dimension along for a ride, driven by the geometry.

Feature-based parametric modeling: Feature-based means that as you create your model, each hole, fillet, chamfer, extrusion etc. is an independent feature that can be edited or deleted.

Bi-directional Associativity: The model and the drawing views are linked. If the model changes, the drawing views will automatically update. And if the dimensions in a drawing view change, the model is updated and the drawing views are updated based on the updated part.

Overview of Model Creation

To get a better idea of the process that you will go through to create a model and its 2D views, refer to the steps outlined below. This is intended as an overview only; not all of the steps are required for every feature that is created.

1. Sketch the geometry.

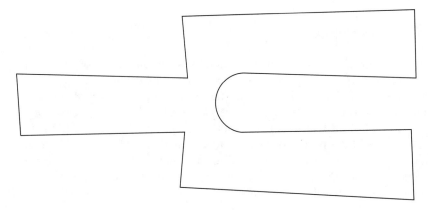

2. Profile the geometry.

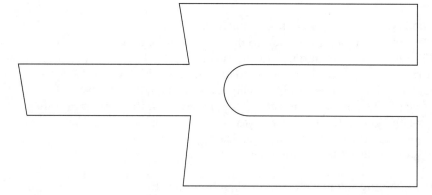

Introduction

3. Add/remove constraints.

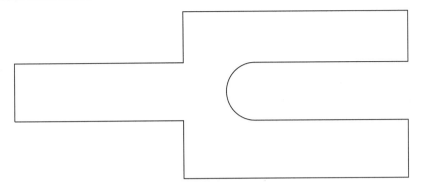

4. Add dimensions.

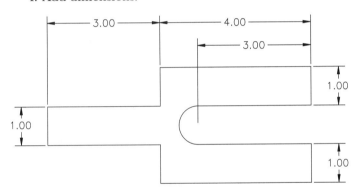

5. Extrude, revolve or sweep the geometry into a solid.

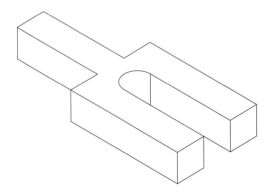

6. Add features (sketched features, holes, fillets, chamfers etc.).

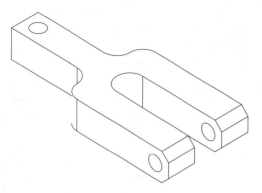

7. Create the 2D views and annotate the drawing.

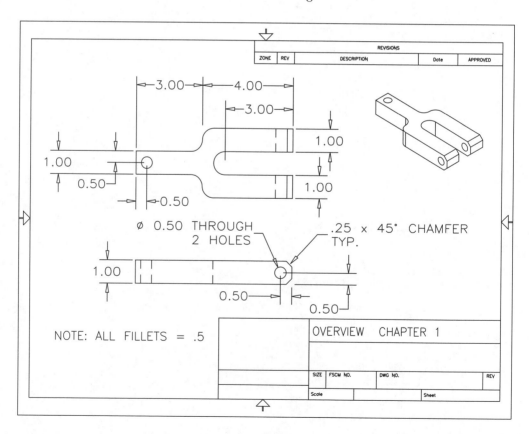

Introduction

Toolbars

The Mechanical Desktop commands can be found in pull-down menus, in four toolbars and through a Desktop browser. This book will focus on the toolbars and the browser. The mcad.mnu uses express toolbars; through the Desktop Application Toolbar, you can change the active toolbar by selecting the operation you want to work with—the mode will change to correspond to the toolbar. Express toolbars require fewer toolbars on the screen. Select from the four icons on the Desktop Application Toolbar to change the current Desktop toolbar. The following four figures show the express tool bars with their respective application toolbars. In this book, the toolbars will be shown in the default vertical orientation, but remember that this is a personal choice, and the toolbars can be placed and oriented anywhere you want.

Part Modeling toolbar

Assembly Modeling toolbar

Another method that can be used to create new parts and to copy, edit and delete parts and features is through the Desktop browser. The Desktop browser shows the history of the file's creation, including features as well as other parts. The capabilities of the browser will be covered throughout the book. Figure I.12 shows the default Mechanical Desktop screen with the browser.

Mechanical Desktop 2.0: Applying Designer and Assembly Modules

Scenes toolbar

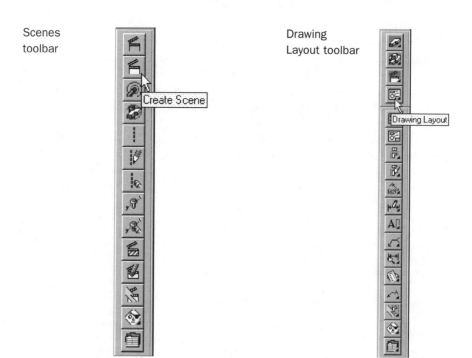

Drawing Layout toolbar

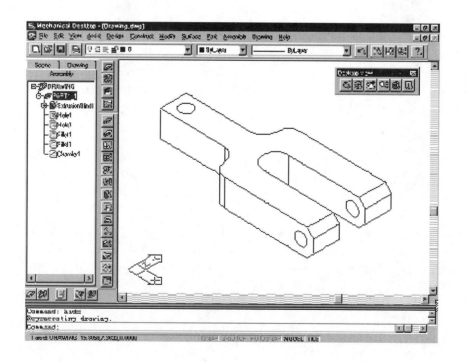

Introduction

Book's Intent

The intent of this book is to focus on the Designer and the Assembly modules in an applied, hands-on environment. This book guides you through the process of generating 3D parametric models, assembling parts together and producing 2D views from them. At the end, Chapter 10 will introduce you to the capabilities of surface modeling.

Each chapter will be broken into specific subjects, which are introduced and then followed by a short tutorial. When a new command is introduced, there is a figure showing the exact location of the icon in the specific toolbar. Read through each subject and then complete its tutorial while at the computer. At the end of each chapter, there are exercises you can complete on your own and review questions to reinforce the topics covered in that chapter.

Special Sections

You will find sections marked:

Notes: Here you will find information that points out specific areas that will help you learn Mechanical Desktop.

Tips: Here you will find information that will assist you in generating better models.

Book Notations

- ⌨ refers to Enter on the keyboard
- Numbers in quotation marks are numbers that need to be typed in.
- Part and model both refer to a Mechanical Desktop part.
- Select refers to a pick by the left mouse button.
- Choose refers to a selection from a menu.
- Desktop browser and browser both refer to the Mechanical Desktop browser, where the history of the file is shown.
- Sketch refers to lines, arcs, circles and polylines drawn to define an outline shape of a feature.

▶ Profile refers to a sketch that has been profiled (analyzed) by Mechanical Desktop.

The Tutorials and Exercises

The best way to learn Mechanical Desktop is to practice the tutorials and exercises on the computer. Each book ships with a CD, and it is recommended that you install the book files on your hard drive. In the MD2BOOK directory, a drawings subdirectory contains subdirectories for each chapter.

Tip: Make a copy of your Mechanical Desktop 2.0 startup icon and change the properties of the "Start in:" directory to "c:\md2book\drawings\chapter??" (where "c" represents the disk drive where the files are located on your system and the "??" represents the chapter you are working on). If you update this startup location as you work through the book, you will reduce the number of selections required to open and save files.

Use your own template when you start a new file. As you go through the book you will find some of the exercises have already been started for you; they can be opened as noted in each exercise. As you go through each exercise, save the files as noted, because they may be required in a future exercise.

With AutoCAD, there are many ways to complete a model, and that also is true with Mechanical Desktop. After completing the exercises as noted, feel free to experiment with different methods.

Value-added Products on CD-ROM

In addition to the exercise drawing files, there are two products on this CD from CIM-LOGIC, Inc.

Educational Philosophy at CIMLOGIC, Inc.

As a founding member of the Autodesk Mechanical Applications Initiative, CIMLOGIC, Inc., is dedicated to the concept of delivering sophisticated mechanical and electrical plug-in software applications in a form usable by a broad base of designers and engineers throughout the world. Its hope and desire is to see computer-aided design and drafting become a commonly used tool by everyone involved in the mechanical engineering field.

Toward this end, CIMLOGIC encourages students and professionals alike to seek out and embrace the wide variety of educational resources currently available for the mechanical design discipline. CIMLOGIC believes that each time an engineer or designer pursues continued training and education, the experience not only contributes to the individual's personal and professional growth, but also adds to the vitality of the entire industry, fostering greater productivity, proficiency and creativity. Whether it is a conference at an industry trade show, a course at an Autodesk Training Center, an Internet university, textbooks or multimedia instruction, today's educational outlets offer a rich array of opportunities, available in formats that can conform to individual needs.

CIMLOGIC Products

There are two CIMLOGIC products on this CD: Desktop Companion for design automation and Toolbox/SM for sheet metal design and unfolding. Both products are fully integrated with Autodesk Mechanical Desktop.

Desktop Companion: Desktop Companion is an easy-to-use, plug-in application that will streamline the design process with Mechanical Desktop.

- Desktop Companion takes industry standards beyond dimensioning to include design standards for model features. Designers can easily comply with industry or company standards—just set the preferences and Desktop Companion does the rest.

- Desktop Companion enhances the design process with functionality for standards-based mechanical design. It saves time creating models with features such as holes, slots, and industry-standard hardware, as well as valuable time searching the Machinery's Handbook because that data is presented in Desktop Companion.

- Part Explorer creates part data and stores it in libraries for projects, product lines, customers, vendors, or any desired category. Part information is stored in a user-defined Microsoft Access compatible database. Parts from libraries can easily be inserted into an assembly or drawing—just point, click and go!

Toolbox/SM for Solids: Toolbox/SM for Solids provides sheet metal design and unfolding tools for use with Autodesk Mechanical Desktop. Toolbox/SM features:

- Tools for creating fully parametric sheet metal models, including bends and punch features
- Flat pattern development—unfold the sheet metal model in a single pick
- Annotation features for full flat pattern details
- Full configuration for setting preferences, layer management, and material data

The Test Drive

You can test drive Desktop Companion and Toolbox/SM for a limited time at no charge. Simply install the product(s) you wish to test drive. When you get to the Authorization Code screen, print out the form and fax it to CIMLOGIC. You will receive an authorization code by return fax. It takes no more than 1/2 hour to get an authorization code. The authorization code lets these fully functional products run for 30 days. Authorization codes are issued by fax only between 8:00 am and 5:00 pm Eastern time. There is no charge for the test drive. There are no POs, no invoices, no obligation. You can pass the CD along to an associate and they can test drive any product.

We Want To Hear From You!

Many of the changes to the look and feel of this new edition were made by way of requests from users of our previous editions. We'd like to hear from you as well! If you have any questions or comments, please contact

>The CADD Team
>c/o Autodesk Press
>3 Columbia Circle
>P.O. Box 15015
>Albany, NY 12212-5015

chapter 1

Sketching, Profiling, Constraining, and Dimensioning

To create a parametric solid, you always start with a 2D profile. After a discussion about specifying the part settings, this chapter will take you through the four steps in generating 2D parametric profiles: drawing a rough 2D sketch of the geometry, profiling (analyzing) the geometry, applying geometric constraints, and finally, adding parametric dimensions.

After completing this chapter, you will be able to:

- Describe how part settings (preferences) affect the creation of a part.
- Sketch the outline of a part.
- Profile a sketch.
- Constrain a sketch.
- Dimension a sketch.

Specifying the Part Settings

As with AutoCAD, certain settings that you specify affect how the system operates and what you see on screen. In Mechanical Desktop, the command for specifying these settings is Edit Preferences (AMPREFS). It can be found on all the toolbars. After you issue the Edit Preferences command (see Figure 1.1), the Desktop Preferences dialog box appears, as shown in Figure 1.2. Each of the five tabs—Part, Assemblies, Surfaces, Drawing, and Desktop—allows changes for specific areas of Mechanical Desktop. In this section we will look at the Part tab. The options offered in the Part tab of the Desktop Preferences dialog box are explained below.

Apply Constraint Rules: If this check box is selected, then when the 2D sketch is profiled, Mechanical Desktop will apply geometric constraints to the sketch. This will help reduce the number of dimensions and constraints required to fully constrain the sketch.

Figure 1.1

Figure 1.2

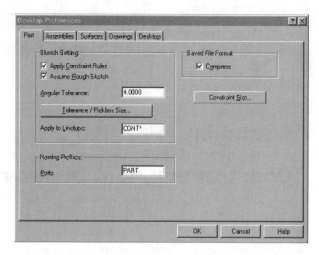

The most common constraints are horizontal, vertical, and tangent. These, along with other constraints, will be covered later in this chapter.

Assume Rough Sketch: If this box is not selected, the sketch will not change after being profiled. When the geometry has been drawn to the exact size, keep this box unchecked. If this check box is selected, the sketch profile will be analyzed on the basis of the angular tolerance and pickbox size. If the sketch's geometry falls within the angular tolerance, the geometry will snap to horizontal or vertical. For example, if two lines are within 4° of parallel, they will be changed and labeled parallel; or if a line is 3° from vertical, it will be changed to vertical.

Angular Tolerance: The value specified in this text box is used when Mechanical Desktop is analyzing the sketch (the default value is 4). For example, any line that is sketched within 4° of horizontal or vertical will be made, respectively, horizontal or vertical. You can change this value to better reflect your geometry.

Tolerance/Pickbox Size...: This button allows you to adjust the size of your pickbox, used in closing a profile. If a sketch has a gap or overlap that is smaller than the pickbox size, it will close the gap. Remember that the pickbox size does not change as you zoom in or out. The further you are zoomed out, the larger the gap/overlap can be.

Sketching, Profiling, Constraining, and Dimensioning

Apply to Linetype: The linetype you select here is the linetype that will be turned into a solid; other linetypes will be used for construction purposes only. Construction lines are discussed further in Chapter 5.

Naming Prefixes: The **Parts** text box in this area lists the default name that each part in a drawing will be given and then sequenced with a number. For example, PART1, PART2. Even if you use this default naming scheme, you can type in a new name when a part is created.

Saved File Format: If the **Compress** check box is selected, the model will be saved in a compressed file. The knowledge of how the part was created will be tightly stored inside the file. The first time a compressed model is opened, it will be uncompressed and will rebuild itself.

Constraint Size...: This button allows you to change the height of the constraint symbols while displaying, editing, or deleting them.

Any changes you make in the part settings are saved in the current drawing and affect geometry created after the changes have been made. If you want these settings to be used for other drawings, set them inside a template file. If as you work with Mechanical Desktop, you are not getting the results that you anticipated, referring back to this section will help you to see whether a setting needs to be adjusted.

Note: Make any changes in the Desktop Preferences dialog box before you profile your sketch.

Sketch the Outline of the Part: Step 1

All 3D models must first start with a 2D sketch of the outline shape of the solid. When deciding what outline to start with, analyze what the finished shape will look like. Look for the shape that describes the part best. When looking for this outline, try to look for a flat face. It is usually easier to work on a flat face than on a curved edge, which can be difficult for new users. However, as you gain modeling experience, reflect back on how the model was created and think about other ways that the model could have been built. Just as with AutoCAD, there is usually more than one way to generate a given model.

When you are working in 2D, you draw the geometry to the exact size. When sketching, you simply draw the geometry so it looks close to the desired shape and size; you do not need to be concerned about exact dimensional values.

Here are some guidelines that will help you generate good sketches.

Tips for sketching:

- **Select an outline that represents the part best.** It is usually easier to work from a flat face.
- **Draw the geometry close to the finished size.** For example, if you want a 2" square, do not draw a 200" square.
- **Create the sketch proportionate to the finished shape.** For instance, to help maintain correct proportions for geometry that is about 10 x 5 in size, you could draw a 10 x 5 rectangle, sketch the geometry inside the rectangle, and then erase the rectangle.
- **Draw the geometry so that it does not overlap.** The geometry should start and end at the same point. See Figure 1.3.
- **Do not allow the geometry to have a gap larger than the pickbox size.** If the gap is smaller than the pickbox size, it will automatically be closed. See Figure 1.4.

Figure 1.3

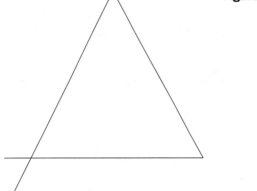

Figure 1.4

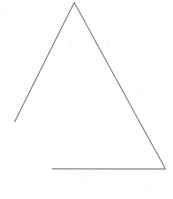

- **Do not draw islands.** For example, if you want a plate with holes in it, first generate the plate. Once it is a solid, then create the holes. See Figure 1.5.
- **Keep the sketches simple.** Leave out fillets and chamfers. They can easily be placed after the sketch is turned into a solid. The simpler the sketch, the fewer the number of constraints and dimensions that will be required to constrain the model.
- **Initially sketch thin areas larger.** Once the sketch is fully constrained, you can go back and change the dimensions to smaller values.

Figure 1.5

 Note: If you are a new user, you may find it helpful to draw the geometry exactly until you feel comfortable with Mechanical Desktop. And some people find it faster to draw the sketch to the exact size, profile it with the Assume Rough Sketch option unchecked, and then apply parametric dimensions, than to draw rough sketches and apply dimensions and constraints. This is a matter of personal preference. This book, however, will draw rough sketches.

Profile the Sketch: Step 2

After the geometry is sketched, you will have Mechanical Desktop analyze it. This is referred to as profiling. Remember, the settings that you specified in the Desktop Preferences dialog box for angular tolerance, applying constraint rules, and assuming a rough sketch will be applied to the geometry in this step.

Select the Profile a Sketch command (AMPROFILE) from the Part Modeling toolbar (see Figure 1.6), select the sketch using any AutoCAD selection technique, and then press ⏎.

After the geometry is profiled, the number of dimensions or constraints required to solve the sketch will be displayed on the command line. If you receive a message:

```
Select edge to close profile: The sketch does not form a closed
profile
```

you can either select an existing edge on the part to close the sketch or press ⏎ to exit the command, use grips to close the sketch, and then profile the sketch. The method for closing an open profile by selecting an existing edge is covered in Chapter 5.

After you have profiled the sketch, a small square with an X through it will appear where the sketch was started. This is referred to as the fixed point, that is, the point to which all geometry will shrink or from which it will grow, depending on the dimensions that are added. You can change this fixed point by issuing the Fix Point command (AMFIXPT) from the Part Modeling toolbar (see Figure 1.7). After issuing the command, select near the point that you want as the new fixed point. You do not need to use object snaps; the fixed point will automatically snap to the nearest endpoint of a line or to the center of an arc or circle. Do not change the fixed point after placing a dimension.

Figure 1.6

Figure 1.7

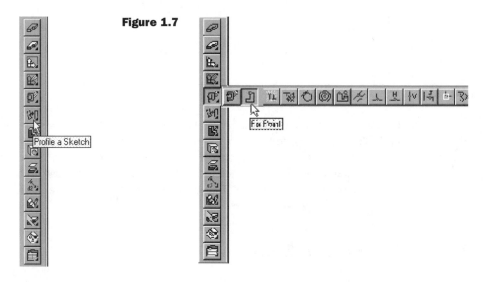

Note: When selecting the sketch to be profiled, you can use any AutoCAD selection method: select individual entities, window, crossing, fence, etc.

Tutorial 1.1—Sketching and Profiling (Bottom Clamp)

1. Start Mechanical Desktop.
2. Start a new drawing.
3. Sketch the geometry as shown in Figure 1.8; approximate size is 5 x 2.
4. Profile the sketch by issuing the Profile a Sketch command (AMPROFILE) and press ⏎.
5. If the geometry was sketched within 4° of horizontal and vertical, it should require seven dimensions or constraints. If your sketch requires more, you will add them in the next section.

Sketching, Profiling, Constraining, and Dimensioning

6. Save the file with the following name:

   \Md2book\Drawings\Chapter1**EX1-1.dwg**

Figure 1.8

Figure 1.9
Finished Bottom Clamp

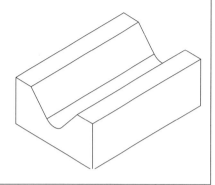

Constrain the Sketch: Step 3

Once a sketch is profiled, Mechanical Desktop automatically analyzes it. If Apply Constraint Rules is checked in the Desktop Preferences dialog box (see Figure 1.2), Mechanical Desktop applies geometrical constraints.

Mechanical Desktop can apply 12 types of constraints. Figure 1.10 shows these constraint types and the letter symbols used to represent them.

Figure 1.10

Constraint	Letter	Example figure
Horizontal	H	Figure 1.11
Vertical	V	Figure 1.12
Perpendicular	L	Figure 1.13
Parallel	P	Figure 1.14
Xvalue	X	Figure 1.15
Yvalue	Y	Figure 1.16
Collinear	C	Figure 1.17
Concentric	N	Figure 1.18
Project	J	Figures 1.19, 1.20
Join	None	Figure 1.21
Radius	R	Figure 1.22
Tangent	T	Figure 1.23

Mechanical Desktop 2.0: Applying Designer and Assembly Modules

Examples for each constraint follow, with the figures on the left showing the sketch before constraints and the figures on the right showing the sketch after the constraints have been applied. For clarity, all the constraints in each example have been removed except those that are discussed in the example.

Figure 1.11
H = Horizontal constraint: Lines are drawn parallel to the X axis.

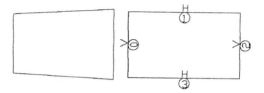

Figure 1.12
V = Vertical constraint: Lines are drawn parallel to the Y axis.

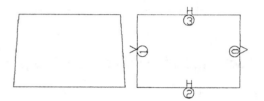

Figure 1.13
L = Perpendicular constraint: Lines will be sketched at 90° to one another; that is, the first line you select will stay and the second will rotate until the angle between them is exactly 90°.

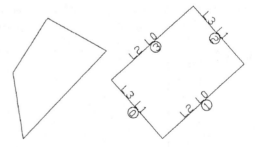

Figure 1.14
P = Parallel constraint: Lines will be sketched exactly parallel to one another; that is, the first line you select will stay and the second will move to become parallel to the first.

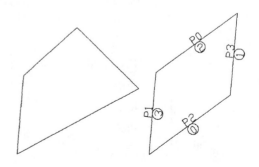

Sketching, Profiling, Constraining, and Dimensioning

Figure 1.15
X = X value constraint: Center points of arcs and/or circles will have the same X coordinate.

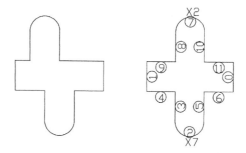

Figure 1.16
Y = Y value constraint: Center points of arcs or circles will have the same Y coordinate.

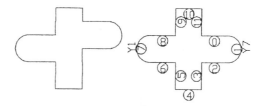

Figure 1.17
C = Collinear constraint: Both lines will line up along a single line; if the first line moves, so will the second.

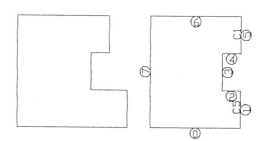

Figure 1.18
N = Concentric constraint: Arcs and/or circles will share the same center point.

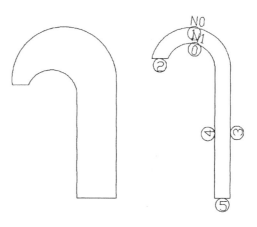

Mechanical Desktop 2.0: Applying Designer and Assembly Modules

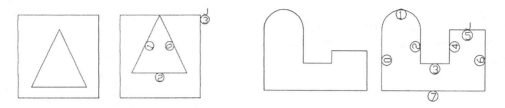

Figures 1.19 and 1.20
J = Project constraint: Using an object snap, you select a point (endpoint, center point). Again using an object snap, you then select a line, an arc, or a circle, which this should touch. Depending on where you select the points, the geometry may move along the X or the Y axis. If this is a problem, place a few dimensions first and then apply the project constraint.

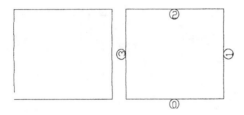

Figure 1.21
Join constraint: The gap between two endpoints of arcs and/or lines will be closed. The Join constraint is applied automatically when the sketch is profiled and the gap is smaller than the AutoCAD pickbox size. There is no symbol for this constraint.

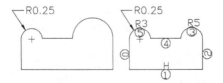

Figure 1.22
R = Radius constraint: Arcs will have the same radius. If this constraint is applied after one of the arcs is dimensioned, the second arc will take on the first arc's radius. If neither arc is dimensioned, both arcs will share the same value. The arc value shared is usually that of the arc selected first, although this is not true every time. For this reason it is better to dimension an arc and place a radial constraint to that arc and then the other.

Figure 1.23
T = Tangent constraint: An arc and a line will become tangent.

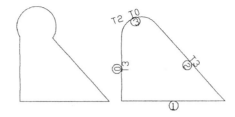

Note: The Project, X value, and Y value constraints are not applied when a sketch is profiled; they need to be added manually.

Show Constraints

After the sketch is profiled, the number of constraints and dimensions that are required to constrain the sketch fully will be displayed at the command line. To see the constraints that were applied, issue the Show Constraints command (AMSHOWCON) from the Part Modeling toolbar; see Figure 1.24. After selecting the command, press **A** and then press ⏎ to show all the constraints on the screen.

You will see numbers enclosed in circles going around the parts. These show the order in which the geometry was created, and they also function as labels for objects. The numbers, then, serve to show relationships between different objects. For example, object 3 is parallel to object 8, or objects 3 and 4 are tangent. Don't worry if the numbers do not go around the part in sequence. The numbers reflect the order in which the geometry was drawn. You will also see letters near the entities in the sketch. The letters tell the types of constraint that were applied (see Figure 1.10).

When you are done viewing the constraints, press ⏎ to return to the command line.

Add Constraints

If you need to add a constraint, select the particular constraint option from the Part Modeling toolbar, as shown in Figure 1.24.

Note: If you click on the first icon, the toolbar will float on the screen until closed. After selecting the constraint option, you will be returned to the drawing, where you can select the geometry to which to apply the constraint. You will stay in this option until you press the ⏎ key, and then you will be given the option at the command line either to select a different constraint to apply or to press ⏎ to exit the command. As the constraints are applied, the number of constraints and dimensions required to solve the sketch will decrease.

Figure 1.24

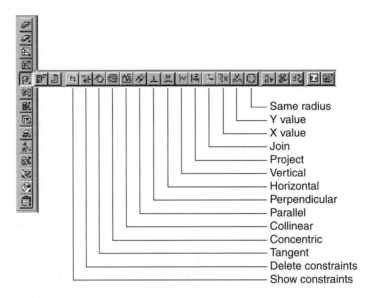

Delete Constraints

To delete a constraint on a sketch, issue the Delete Constraints command (AMDELCON) command from the Part Modeling toolbar, as shown in Figure 1.24. After you have issued the command, the constraint information will appear if there is only one sketch. Otherwise, you will be prompted to select a sketch. If the constraint symbols appear too small or too large, press S and press ⏎ to adjust the size of the constraints. To delete a constraint, simply select the constraint symbol located on the given line, arc, or circle. After you select the constraint symbol, the number of constraints and dimensions required to constrain the model will be increased. To delete another constraint, select that constraint. To exit the command, press ⏎.

Tutorial 1.1 (continued)—Constraining (Bottom Clamp)

1. If the current file is not \Md2book\Drawings\Chapter1\EX1-1.dwg, then open it now.

2. Use the Show Constraints command (AMSHOWCON) to show the constraints that were automatically applied. Issue the command and then press A to display all constraints. Your drawing should look like Figure 1.25. If you are missing a constraint, you can add it in step 3. When you are done looking at the constraints, press ⏎ to exit the command.

Sketching, Profiling, Constraining, and Dimensioning

Figure 1.25

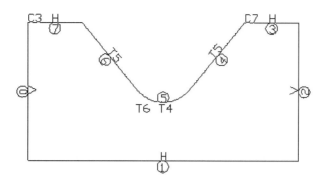

3. Add missing constraints as needed, issuing the necessary command(s) from the 2D Constraints toolbar.
4. Save the file.

Dimension the Sketch Profile: Step 4

The last step is to dimension the sketch profile. The dimensions you place will control the size of the model and will also appear in the drawing views when they are generated. Don't be overly concerned about where the dimensions are placed, because usually, they are not exactly where they need to be when the drawing views are finally laid out. They can easily be relocated after the drawing views have been created. Try to avoid having the extension lines go through the part, as this will require more clean-up. With experience you will place the dimensions in better locations, requiring less cleanup when the drawing views are created.

All parametric dimensions are created with a single command, Add Dimension (AMPAR-DIM), from the Part Modeling toolbar (see Figure 1.26). There is no need to use object snaps when placing these dimensions. When you select a line, it will snap to the nearest endpoint; when an arc or circle is selected, it will snap to its center point.

There are two techniques for dimensioning a line: You can select near two endpoints and then select a location for the dimension. Or if you want to dimension a single line, select the line anywhere and then select a location for that dimension.

To create an angular dimension, select near the midpoint of the two lines and then select a point for the dimension location.

To dimension an arc or a circle, select on its circumference and then select a location outside it. When an arc is dimensioned, the result is a radius dimension; when a circle is dimensioned, the result is a diameter dimension.

If the wrong dimension type appears after you select a placement point, you can change it at the command prompt by typing in the capital letter of the constraint type you want. The following string shows what the command prompt will look like:

```
Undo/Hor/Ver/Align/Par/aNgle/Ord/Diameter/pLace/Enter  dimension
value <5.0000>:
```

For example, if a vertical dimension appears and you want an aligned dimension, press **A** and then press [⏎Enter] to change its type. Once the correct dimension type is on the screen, you can either press [⏎Enter] to accept the value in the command line or type in a new value and press [⏎Enter].

When inputting values, type in the exact value; do **not** round up or down. The accuracy shown in the dimension is from the current dimension style. Mechanical Desktop models are accurate to six decimal places; for example, 1.0625 is more accurate than 1.06. When placing dimensions, it is recommended that you place the smallest dimensions first; this will prevent the geometry from flipping in the wrong direction.

As the sketch is constrained and dimensioned, the number of constraints and dimensions required will decrease until the sketch profile is fully constrained. With Mechanical Desktop it is not required that a sketch be fully solved. In Chapter 2 you will learn how to go back and edit a sketch profile after it has been made into a solid or feature.

Figure 1.27, together with the following explanations, shows how to create linear, angular, and radial dimensions.

Figure 1.26

Sketching, Profiling, Constraining, and Dimensioning

Figure 1.27

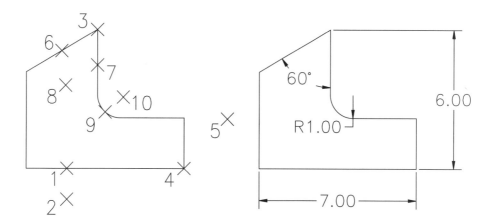

Linear Dimensions

Select a line and then a placement point (points 1 and 2). Or select near two different endpoints and then select a placement point (points 3, 4, and 5).

Angular Dimensions

Select near the midpoints of the two lines you want dimensioned and then select a placement point (points 6, 7, and 8). Or if a linear dimension appears, press **N** at the command line to get an angle dimension.

Radial Dimensions

Select the arc or circle to dimension and then select a placement point (points 9 and 10).

Ordinate Dimensions

To create an ordinate dimension, first select near the point on the sketch that you want to be the zero point and then select a location point. A linear dimension will appear. Then press **O** and then press [Enter]. The dimension will change to an ordinate dimension with a value of zero. Any dimension placed after that will be an ordinate dimension related to the zero dimension. If placing an ordinate dimension based on a zero dimension different from the one the last ordinate dimension was based on, select its extension line for the first selection and then select the point to dimension. You will need to place a separate zero ordinate dimension in both the X and the Y directions.

Changing Dimensions

To change a dimension's value, issue the Change Dimension command (AMMODDIM) from the Part Modeling toolbar (see Figure 1.28), and then select the dimension you want to change. Its current value will be displayed on the command line. Type in a new value and press [Enter]. To change another dimension's value, select the dimension, or to exit the command, press [Enter]. After the command is exited, the sketch will automatically be updated to reflect this new value.

15

Figure 1.28

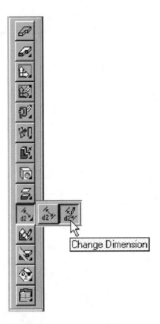

Notes: When you input dimensional values, it is recommended that you use decimal values, not fractions. Fractions are interpreted literally. For example, 2-1/2 is interpreted as 2 minus 1/2. If you must use fractions, use a + sign, for example, 2+1/2.

Dimensions are created with the current dimension style (number of decimal places, text height, etc.).

You do not have to fully solve (dimension or constrain) your sketch. Mechanical Desktop permits the use of underconstrained models. As you go through this book, you will learn how to return to a sketch profile or model and add or delete dimensions and constraints.

Tips: When typing in values, type in the exact values; do not round up or down. The accuracy shown in the dimension is from the current dimension style. Mechanical Desktop models are accurate to six decimal places; for example, 1.0625 is more accurate than 1.06.

Place the smallest dimensions first. This will prevent the geometry from flipping in the wrong direction.

When creating profiles with thin areas, it may help to draw them larger. Once the sketch is fully constrained, you can go back and change the dimensions to smaller values.

Sketching, Profiling, Constraining, and Dimensioning

Tutorial 1.1 (continued)—Dimensioning (Bottom Clamp)

1. If the current file is not \Md2book\Drawings\Chapter1\EX1-1.dwg, then open it now.

2. To add dimensions, issue the Add Dimension command (AMPARDIM). Dimension the profile as shown in Figure 1.29. Your dimensions may differ, depending on the settings in your current dimension style.

Figure 1.29

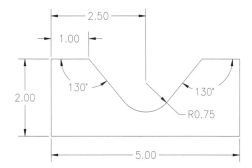

3. To change a dimension, issue the Change Dimension command (AMMODDIM). Select the 2 .00 dimension and change it to 2.5, and change the .75 radius to .625. Press ⏎ to exit the command. The completed profile should look like Figure 1.30.

Figure 1.30

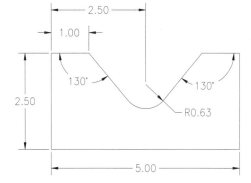

Mechanical Desktop 2.0: Applying Designer and Assembly Modules

Tutorial 1.2—Sketching, Profiling, Constraining, and Dimensioning (Top Clamp)

Figure 1.31

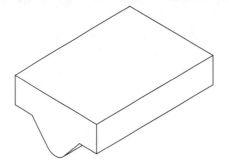

1. Start a new drawing.
2. Sketch the geometry as shown in Figure 1.32. The approximate size is 5 x 4.

Figure 1.32

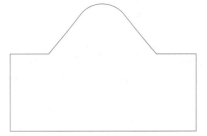

3. Profile the geometry. The sketch should require at least seven dimensions or constraints.
4. Show the constraints. Your sketch should look like Figure 1.33.

Figure 1.33

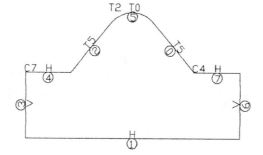

5. Add constraints if required.

6. Add the dimensions shown in Figure 1.34.

Figure 1.34

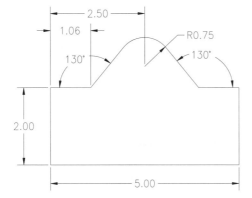

7. Change the 2.00 dimension to 1.5, the .75 radius to .625, and the 1.00 dimension to 1.0625. The completed profile should look like Figure 1.35.

Figure 1.35

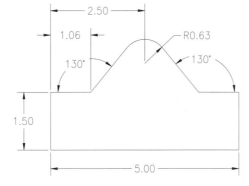

Mechanical Desktop 2.0: Applying Designer and Assembly Modules

Exercises

The following exercises are intended to challenge you by providing problems that are open-ended. As with the kinds of problems you'll encounter in work situations, there are multiple ways to arrive at the intended solution. In the end, your solution should match the drawing shown. For each of the following exercises, start a new drawing. When you have finished the drawing, save the file using the name supplied in the exercise.

Exercise 1.1—Bracket

Save the finished drawing (Figure 1.36) as \Md2book\Drawings\Chapter1\Bracket.dwg.

Figure 1.36

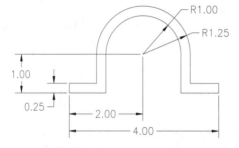

Exercise 1.2—Guide

Save the finished drawing (Figure 1.37) as \Md2book\Drawings\Chapter1\Guide.dwg. (Hint: Use Radial, X Value, and Y Value constraints.)

Figure 1.37

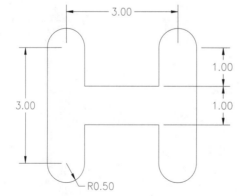

Sketching, Profiling, Constraining, and Dimensioning

Exercise 1.3—Foot

Save the finished drawing (Figure 1.38) as \Md2book\Drawings\Chapter1\Foot.dwg.

Figure 1.38

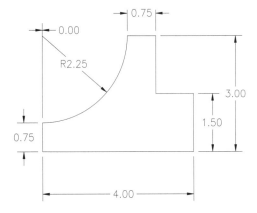

Exercise 1.4—Lamp (quarter profile of a lamp)

Save the finished drawing (Figure 1.39) as \Md2book\Drawings\Chapter1\Lamp.dwg

Figure 1.39

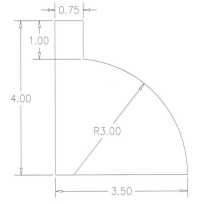

Chapter 1 Review Questions

1. List four items that can be set in the Desktop Preferences dialog box.
2. How big a gap can you have in a sketch?
3. A sketch profile does not need to be fully constrained. T or F?
4. If a sketch is drawn to exact size, does it need to be profiled? What setting needs to be checked in the Desktop Preferences dialog box?
5. When drawing a sketch, it is OK to have islands in it. T or F?
6. Can a constraint be removed? If so, how?
7. When an angled dimension is required and a horizontal dimension appears, can this be changed to an angle dimension? If so, how?
8. When an arc is dimensioned, is the default a radius or a diameter dimension?
9. To how many decimal places is a Mechanical Desktop model accurate?
10. After a sketch is fully constrained, a dimension's value cannot be changed. T or F?

chapter 2

Viewing, Extruding, Revolving, Sweeping, and Editing Parts

After you have drawn, profiled, constrained, and dimensioned the sketch, your next step is to turn that sketch into a 3D part. This chapter takes you through the options for viewing a part from different viewpoints, as well as the options for creating, extruding, revolving, sweeping, and editing parts.

After completing this chapter, you will be able to:

- View a model from different viewpoints.
- Extrude a profile into a part.
- Revolve a profile into a part.
- Sweep a profile into a part.
- Edit features of a part.

Viewing a Model from Different Viewpoints

Up to now you have worked with 2D sketches in the XY plane, as you have done in 2D AutoCAD. The next step is to give the 2D sketch depth in the Z plane. To see 3D geometry, you must be able to view the model from different viewpoints.

Options for Viewing the Model

AutoCAD has many options for viewing the model from different perspectives. One option is to use predefined model views. With Mechanical Desktop, you have two other options for viewing the model: number keys and dynamic rotation.

Model Views

To use predefined views, select **Model Views** from the View pull-down menu and select the desired option from the cascading menu.

Number Keys

When you choose the number keys option for viewing the model, you use the number keys on the keyboard to change the screen layout and viewpoint.

The 1, 2, 3, and 4 keys split the screen into corresponding numbers of viewports, each with different viewpoints:

Key	Viewport(s)
1	One (taken from the last active viewport)
2	Two (top and isometric views)
3	Three (top, front, and isometric views)
4	Four (top, front, side, and isometric views)

The following keys provide the following views:

Key	View
5	Top
6	Front
7	Side
8	Southeast isometric
88	Southwest isometric
9	Plan view of the current sketch plane

To use number keys, choose the desired number at the command prompt and press Enter.

Dynamic Rotation

The second option for viewing the model in Mechanical Desktop is to rotate the model dynamically. There are six ways to dynamically rotate the model. You can access these six ways on the Desktop View toolbar (see Figure 2.1). Each way is denoted by an icon.

Figure 2.1
Desktop View toolbar

The Desktop View toolbar has six icons, each with fly-out icons, that are described below. The first icon and its fly-out icons on the Desktop View toolbar allow you (1) rotate the model about its geo-center, (2) rotate the model around a selected point, (3) use the AutoCAD DVIEW command, or (4) rotate the model incrementally using the keyboard. To rotate a model dynamically using this icon, first select the top icon, then select a point with the left mouse button. Hold the button down while moving the mouse. The model rotates in the mouse's direction. When you release the mouse button, the model stops rotating. To return to the original position press the Esc key. To accept the current position, press ⏎.

The second icon and its fly-out icons on the Desktop View toolbar let you view the model in shaded, transparent, wireframe, or hidden mode. The sixth, seventh and eighth icons on this fly-out allow you to select the objects to rotate, apply material value such as transparency, roughness, color, reflection and ambience. This is only applicable if Apply materials is turned on in the set 3D graphic preferences and the last icon set 3D graphic preferences, respectively. Models can be edited in any of these modes. When selected, the top icon toggles between shaded and wireframe modes.

The third icon and its fly-out icon have panning commands.

The fourth icon and its fly-out icons have zooming commands.

The fifth has preset views.

The sixth icon allows you to save your screen configuration and replay it. After getting the screen in the orientation that you want to save, select from the fly-out the red number 1, 2 or 3. To change the screen back to this configuration, select the black corresponding number from the fly-out. The last four icons on the fly-out will split the screen into 1, 2, 3 or 4 viewports respectively.

Tutorial 2.1—Viewing a Model

1. Open the file \Md2book\Drawings\Chapter2\Ex2-1.dwg.

2. Use the number keys to change view configurations.

3. When you feel comfortable changing views, return to one viewport by pressing the 1 key, then pressing [↵].

4. To return to plan view, press the 9 key, and then press [↵].

5. Try the dynamic rotation options by selecting the Dynamic Rotation icon from the Desktop View toolbar. Hold the mouse button down while rotating the model.

6. Alternate between shaded, transparent, wireframe, and hidden modes, and rotate the model dynamically in each. When you feel comfortable rotating the model, proceed to the next section.

Tips: You can edit models in shaded, transparent, wireframe, or hidden mode.

If you choose the Dynamic Rotation icon and no objects appear return to the Select Objects icon (the sixth icon in the third column) and select the model you wish to rotate.

To remove objects from the selection set, hold the Shift key and select the objects for removal.

Using the XY Orientation in 3D

Now that you have sketched, profiled, and constrained the model, and viewed and rotated it from different perspectives, you are ready to create a 3D part. When you work in 3D, it is important that the UCSICON (icon with horizontal and vertical arrows in the lower left corner of the screen) be turned on. Many 2D AutoCAD users turn this icon off because the XY orientation does not change. When working in 3D, you will continually be changing the location of the XY plane and will use the UCSICON as a visual aid. If this icon is off, use the On option of the UCSICON command.

The W just below the Y, which refers to the World Coordinate System, is the default setting. If the W is not below the Y, you are in a user-defined coordinate system (UCS), which means the UCS has been moved. The UCS and how it relates to Mechanical Desktop are covered in more detail in Chapter 3.

Viewing, Extruding, Revolving, Sweeping, and Editing Parts

Extruding the Profile

The first method for creating a part is to extrude the profile, which gives the profile depth along the Z axis. Before extruding, it is helpful to be in an isometric view, because Mechanical Desktop shows you the direction of the extrusion with an arrow (negative or positive). If you are viewing the profile from directly above, and the arrow is coming toward you, you will see a circle with a dot in its center; if it is going away from you, you will see a circle with a cross in it.

To extrude a profile, select the Extrude command (AMEXTRUDE) on the Part Modeling toolbar (see Figure 2.2). After you issue the command, the Extrusion Feature dialog box appears (see Figure 2.3).

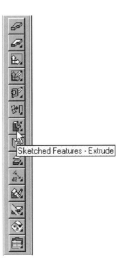

Figure 2.2
Extrude command

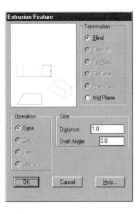

Figure 2.3
Extrusion Feature dialog box

The Extrusion Feature dialog box has three sections: Operation, Termination, and Size.

Operation

In the Operation section of the Extrusion Feature dialog box, you specify if this is the first profile you will extrude (base feature) or if you will extrude by adding or removing material to or from an existing part. This section has four options to choose from. The Base option is the default for the first extrusion. The Cut, Join, and Intersect options are grayed out until after a base feature exists.

Option	Function
B<u>a</u>se	Serves as the default setting for the first extrusion. Creates a base part to or from which other operations add or remove material.
<u>C</u>ut	Determines when the extrusion will remove material from the part.
<u>J</u>oin	Specifies when the extrusion will add material to the part.
<u>I</u>ntersect	Identifies when the extrusion will keep what is common to the first part and the second extrusion.

The Cut, Join, and Intersect options gray out until a base part exists.

Termination

The Termination section of the Extrusion Feature dialog box, which has the following six options, determines the extent of the extrusion. There are six options to choose from. Like the Operation section, this section has options that are grayed out until a base part exists. These items that are grayed out will be are covered in Chapter 4.

Option	Function
<u>B</u>lind	Extrudes a specific distance in the positive or negative Z direction.
<u>T</u>hrough	Goes all the way through the part in one direction.
To <u>P</u>lane	Continues until the profile reaches a specific plane that is flat.
To <u>F</u>ace	Continues until the profile reaches a specific face that is contoured.
F<u>r</u>om To	Starts at one plane/face and stops at another.
Mi<u>d</u> Plane	Goes equal distances in the negative and positive directions. For example, if the extrusion distance is 2", the extrusion goes 1" in both the negative and positive Z directions.

Size

In the Size section of the Extrusion Feature dialog box, you specify the extrusion distance and a draft angle.

Option	Function
Distance	Type the value the profile should be when extruded.
Draft Angle	Type the draft angle. A 0 draft angle extrudes the profile straight outward, a negative number tapers the profile inward, and a positive number expands the profile outward.

Tutorial 2.2—Extruding the Profile

1. Start a new drawing
2. Draw a circle.
3. Profile the circle.
4. Add a 2" diameter dimension to the circle.
5. Change to an isometric view using any technique. Your drawing should resemble Figure 2.4, which was created with the 8 key.

Figure 2.4

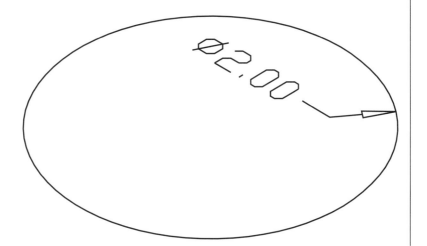

6. Issue the Extrude command (AMEXTRUDE). The Extrusion Feature dialog box appears.
7. Input the following data (see Figure 2.5):

    Operation = B<u>a</u>se (default setting)

    Termination = <u>B</u>lind (default setting)

    Size

    Di<u>s</u>tance = 2

    Draft An<u>g</u>le = –5

8. Select OK. A prompt to accept or flip the extrusion direction appears.
9. To change the extrusion direction, press F and then press [↵]. To return to the original extrusion direction, again press F and then press [↵].
10. Press [↵] to extrude the circle. Your drawing should resemble Figure 2.6.
11. Save your drawing as \Md2book\Drawings\Chapter2\Ex2-2.

Figure 2.5

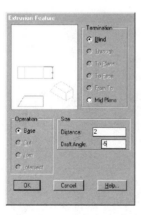

Figure 2.6

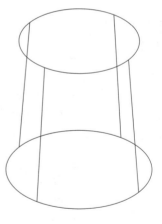

Note: When you convert a profile to a part, all dimensions disappear. For more information on editing parts or features, see "Editing 3D Parts" later in this chapter.

Viewing, Extruding, Revolving, Sweeping, and Editing Parts

Tutorial 2.3—Extruding the Profile Mid-Plane

1. Open the file \Md2book\Drawings\Chapter2\EX2-3.dwg.

2. Issue the Extrude command (AMEXTRUDE). The Extrusion Feature dialog box appears. You do not need to profile or add dimensions, because those steps have already been taken.

3. Input the following data (see Figure 2.7):

    Operation = B<u>a</u>se (default setting)

    Termination = Mi<u>d</u> Plane

    Size

    D<u>i</u>stance = 2

    Draft Angle = –5

4. Select OK. Your drawing should resemble Figure 2.8. As you can see, the profile extrudes equally in the positive and negative directions. As a result, you are not prompted to change the extrusion direction.

5. <u>S</u>ave the file.

Figure 2.7

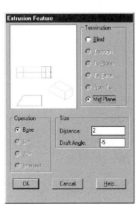

Figure 2.8

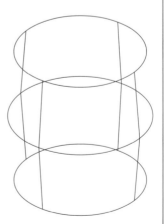

Note: To extend the draft angle from the part, give the draft angle a positive number.

Revolving the Profile

The second method for creating a part is to revolve a profile around a straight edge or axis. In this process, you take the steps you did to extrude the profile (sketch, profile, constrain, and dimension), then you select the Revolve command (AMREVOLVE) from the Part Modeling toolbar (see Figure 2.9).

The Revolution Feature dialog box appears (see Figure 2.10).

Figure 2.9
Revolve command

Figure 2.10
Revolution Feature dialog box

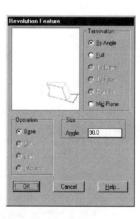

Like the Extrusion Feature dialog box, the Revolution Feature dialog box has three sections: Operation, Termination, and Size.

Operation

In the Operation section of the Revolution Feature dialog box, you specify if this is the first profile you will revolve or if you will revolve the profile by adding or removing material from an existing part. You have four options to choose from: Base, Cut, Join, and Intersect. The Base option is the default setting if this is the first feature created. The Cut, Join, and Intersect options are grayed out until a base part is created.

Viewing, Extruding, Revolving, Sweeping, and Editing Parts

Option	Function
B<u>a</u>se	Serves as the default setting for the first revolve. Creates a base part to or from which other operations add or remove material.
<u>C</u>ut	Determines when the revolve will remove material from the part.
<u>J</u>oin	Specifies when the revolve will add material to the part.
<u>I</u>ntersect	Identifies material to remain, common to the part and the new feature.

The Cut, Join, and Intersect options gray out until you create a part to or from which to add or subtract material.

Termination

In this section, you specify the degrees of revolve. There are six options to choose from. As in the Operation section of the dialog box, there are some options that are grayed out until a base part exists. These items that are grayed out will be are covered in Chapter 4.

Option	Function
<u>B</u>y Angle	Revolves the profile a specific number of degrees in the positive or negative Z direction. A prompt directs you to choose the direction in which to revolve.
<u>F</u>ull	Revolves the profile 360°.
To <u>P</u>lane	Revolves the profile until it reaches a specific plane, or work plane, which is flat.
<u>T</u>o Face	Revolves the profile until it reaches a specific face, which is contoured.
F<u>r</u>om To	Starts the revolve at one plane/face and stops it at another.
Mi<u>d</u> Plane	Revolves the profile equally in the negative and positive directions. For example, if the revolve angle is 30°, the profile revolves 15° in both the negative and positive directions.

Size

In this section, you denote the number of degrees the revolve will travel. There is not an option for draft angle with revolve.

Option	Function
Angle	Type the degrees to which the profile should revolve.

After inputting all data in the Revolution Feature dialog box, select OK. Mechanical Desktop prompts you to select a straight edge around which to revolve the profile. (Chapter 5 discusses constructions geometry, which result from revolving a profile around an axis that is not on a given profile.) Select the edge. If you are revolving by an angle, an arrow appears asking you to accept or flip the default direction. Press F and then press to flip the revolve direction. To accept the current direction, press to revolve the profile and exit the command.

Tutorial 2.4—Revolving the Profile By Angle

1. Start a new drawing of a triangle as shown in Figure 2.11.
2. Draw a triangle as shown in Figure 2.11.
3. Profile the triangle.
4. Add dimensions to the triangle as shown in Figure 2.12.
5. Change to an isometric view.

Figure 2.11

Figure 2.12

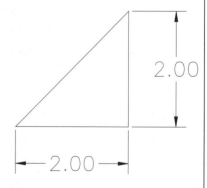

Viewing, Extruding, Revolving, Sweeping, and Editing Parts

6. Issue the Revolve command (AMREVOLVE). The Revolution Feature dialog box appears (see Figure 2.13).

7. Input the following data

    Operation = B<u>a</u>se (default setting)

    Termination = <u>B</u>y Angle (default setting)

    Size

    A<u>n</u>gle = 180

8. Select OK.

9. Select the triangle's vertical line. A prompt asking you to accept or flip the revolution direction appears.

10. To change the revolution direction, press F and then press [⏎]. To return to the original revolution direction, press F and then press [⏎] again to change to the original extrusion direction.

11. Press [⏎] to revolve the triangle. Your drawing should resemble Figure 2.14.

12. Save your drawing as \Md2book\Drawings\Chapter2\EX2-4.

Figure 2.13

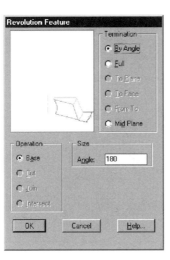

Figure 2.14

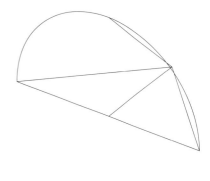

Tutorial 2.5—Revolving the Profile Fully

1. Open the file `\Md2book\Drawings\Chapter2\EX2-5`.

2. Issue the Revolve command (AMREVOLVE). You do not need to profile or add dimensions, because these steps have already been taken. The Revolution Feature dialog box appears.

3. Input the following data:

    Operation = B<u>a</u>se (default setting)

    Termination = <u>F</u>ull

   Note that the Size section grays out. This is because the geometry will revolve 360°.

4. Select OK.

5. Select the vertical line. Your drawing should resemble Figure 2.15.

6. <u>S</u>ave the file.

Figure 2.15

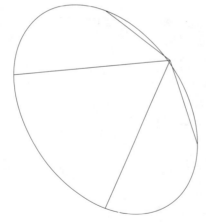

Viewing, Extruding, Revolving, Sweeping, and Editing Parts

Tutorial 2.6—Revolving the Profile Mid-Plane

1. Open the file \Md2book\Drawings\Chapter2\EX2-6.

2. Issue the Revolve command (AMREVOLVE). You do not need to profile or add dimensions, because these steps have already been taken. The Revolution Feature dialog box appears.

3. Input the following data:

    Operation = B<u>a</u>se (default setting)

    Termination = Mi<u>d</u> Plane

    Size

       A<u>n</u>gle: 90

4. Select OK.

5. Select the vertical line. Your drawing should resemble Figure 2.16, which shows the part with lines hidden.

6. <u>S</u>ave the file.

Figure 2.16

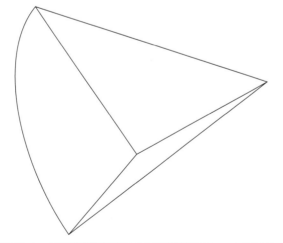

Sweeping the Profile

The third method for creating a part is sweeping. In this section you will learn how to sweep a profile along a 2D path. A sweep is like an extrusion, except that a sweep follows a path. In contrast, an extrusion extrudes the profile in one direction only, in the positive or negative Z axis. Before performing a sweep, you must complete six steps: create the path, constrain the path, create a work plane, sketch a profile, constrain the profile, and sweep the profile along the path.

Create a Path: Step 1

In this step, you use lines, arcs, or polylines to draw a path along which to sweep the profile. This path can be open or closed, but it must lie in one plane.

1. Select the Path (AMPATH) command from the Part Modeling toolbar (see Figure 2.17).

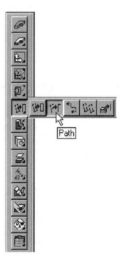

Figure 2.17
Path command

2. Select the geometry. Mechanical Desktop analyzes the sketch like a profile, then Mechanical Desktop prompts you to select a start point from which to draw the path.

3. Choose a point near the geometry (do not use object snaps). The number of constraints and or dimensions required to constrain the path appears on the command line.

Constrain the Path: Step 2

In this step, you add constraints and dimensions to the path, as you did with profiles. You do not need to constrain the path.

Create Work Plane: Step 3

The third step in sweeping is to create a work plane in which to draw the profile. A work plane is like the UCS, but it is tied to the part. (Work planes are detailed in Chapter 3.) For sweeping to work, the work plane must be normal (perpendicular) to the path's start point. Normal is 90° from a given point. For example, if you stand a pencil straight upright on a desk, the pencil is normal to the desk.

Why not rotate the UCS 90° and place its origin at the path's start point? Because the UCS is not tied to the geometry. If you moved the UCS normal to the end of the path, then rotated the path, the UCS would not follow the path. In contrast, the work plane is tied to the geometry. Therefore, the work plane moves when the path moves.

To create a work plane:

1. Issue the Work Plane command (AMWORKPLN) on the Part Modeling toolbar (see Figure 2.18). The Work Plane Feature dialog box appears (see Figure 2.19).
2. In the 1st Modifier section of the dialog box, select Sweep Profile.
3. Below the 1st Modifier section, choose Create Sketch Plane.
4. Select OK. A work plane appears at the path's start point. Mechanical Desktop then prompts you to rotate the XY orientation. Three small lines, called the work point, will appear at the end of the line. You use the work point when dimensioning the profile to the path.

Figure 2.18
Work Plane command

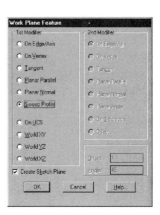

Figure 2.19
Work Plane Feature dialog box

Draw the Profile: Step 4

The fourth step is to draw the profile to be swept along the path. This profile is the same as a profile you would extrude or revolve.

Constrain the Profile: Step 5

The fifth step is to constrain and dimension the profile, then constrain it to the end of the path. You can use the work point to constrain the profile to the end of the path. When you select the work point, it snaps to the intersection of the three lines.

When you dimension a base feature (the first feature), the command prompt does not reflect the fact that profile requires two more dimensions to constrain it to the end of the path. For example, when you sweep a circle along a path, the command prompt reflects that the circle requires only one dimension. While the prompt says "one," it accepts the two dimensions needed to constrain the profile in the X and Y axes.

Sweep the Profile: Step 6

The final step in this process is to sweep the profile along the path. Issue the Sweep command (AMSWEEP) on the Part Modeling toolbar (see Figure 2.20). The dialog box appears. You must set four options: Operation, Termination, Body Type, and Size.

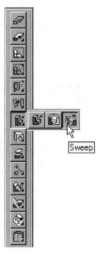

Figure 2.20
Sweep command

Operation

In the Operation section of the dialog box, you specify if this is the first part you will sweep or if you will sweep by adding or removing material from an existing part. You have four options: The Base option is the default setting for the first feature. The Cut, Join, and Intersect options gray out until the base part exists.

Viewing, Extruding, Revolving, Sweeping, and Editing Parts

Option	Function
Base	Serves as the default setting for the first sweep. Creates a base part to or from which other operations add or remove material.
Cut	Removes material from the part.
Join	Adds material to the part.
Intersect	Keeps what is common to the first and second parts.

Termination

This section allows you to identify where you want the profile to start and end. You have three options:

Option	Function
Path Only	Causes the profile to follow the path from its start point to its end point.
To Face	Begins the sweep at the path's start point but stops the sweep at a specified face.
From To	Starts the sweep at a specified plane/face but stops it at a different plane/face.

In this section, two options are grayed out until the base part exists. The grayed items are covered in Chapter 4.

Body Type

In this section, you determine the way in which the profile should travel along the path. You have two options:

Option	Renders the profile
Normal	90° from the path (see Figure 2.21)
Parallel	Parallel to the profile in the start position as it is swept along the path (see Figure 2.21)

Figure 2.21
Comparing normal and parallel

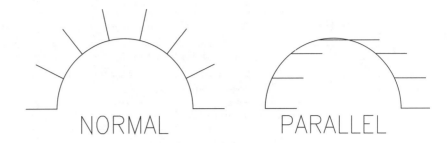

Size

In this section, you specify the draft angle. In the Draft Angle box, type the draft angle. A 0 draft angle extrudes the profile without enlarging or reducing the profile, a negative number drafts the geometry inward, and a positive number drafts the geometry outward.

After making changes to the dialog box, select OK. The profile sweeps along the path as specified.

Tutorial 2.7—Sweeping an Open Path

1. Start a new drawing.
2. Draw a line and an arc as shown in Figure 2.22.

Figure 2.22

3. Issue the Path command (AMPATH).
4. Select the geometry.
5. Select near the left end of the horizontal line.
6. Add the dimensions shown in Figure 2.23.

Figure 2.23

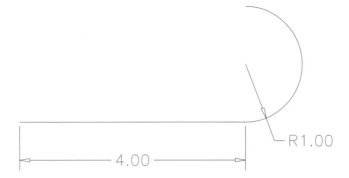

7. Use the 8 key to change to an isometric view.
8. Issue the Work Plane command (AMWORKPLN). The dialog box appears.
9. In the 1st Modifier section, select Sweep Profile.
10. Below the 1st Modifier section, check Create Sketch Plane.
11. Select OK.
12. At the Rotate/Z-flip/<Accept>: prompt, press [⏎] to accept the default XY orientation.
13. Draw a circle near the end of the line, but do not use object snaps.
14. Profile the circle. The command prompt indicates that you require only one constraint or dimension.
15. Add three dimensions as shown in Figure 2.24. Although the prompt requests one dimension, you add three because the profile is the base feature. The profile command does not know that this profile will be swept. (A circle only requires one constraint.) Create the two 0 dimensions, which are for locating the circle to the work point, by selecting the circle and the work point and placing the dimension in the horizontal and vertical positions. A concentric constraint could have been used instead of the two 0 dimensions.

Figure 2.24

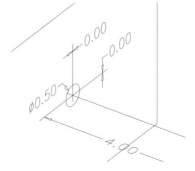

16. Issue the Sweep command (AMSWEEP). The dialog box appears.
17. Select OK. The profile sweeps along the path. Your model should resemble Figure 2.25.

Figure 2.25

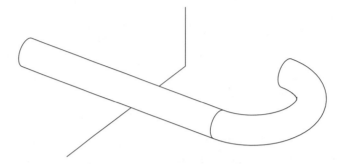

18. Save your file as \Md2book\Drawings\Chapter2\EX2-7.dwg.

Tutorial 2.8—Sweeping a Closed Path

1. Start a new drawing
2. Draw a slot as shown in Figure 2.26.
3. Issue the Path command (AMPATH).
4. Select the geometry.
5. Select the left side of the lower horizontal line.
6. Dimension the slot as shown in Figure 2.26.

Figure 2.26

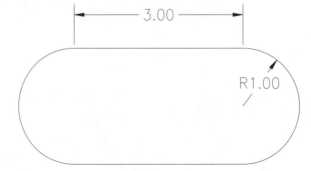

Viewing, Extruding, Revolving, Sweeping, and Editing Parts

7. Change to an isometric view using the 8 key.
8. Issue the AMWORKPLN command. The dialog box appears.
9. In the 1st Modifier section, select Sweep Profile.
10. Below the 1st Modifier section, check Create Sketch Plane.
11. Select OK.
12. Press R and press ⎯⏎⎯ twice to rotate the XY icon so that the positive X is pointing into the screen.
13. Draw a square near the left end of the lower horizontal line as shown in Figure 2.27.
14. Profile the square. The command prompt indicates that you require only two constraints or dimensions.
15. Add four dimensions as shown in Figure 2.27. (You add two dimensions for the square and two to locate the square to the slot.)
16. Issue the Sweep command (AMSWEEP). The dialog box appears.
17. Select OK to accept the defaults. The profile sweeps along the path. Your drawing should resemble Figure 2.28 shown with lines hidden.
18. Save your file as \Md2book\Drawings\Chapter2\EX2-8.dwg.

Figure 2.27

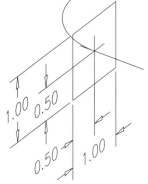

Figure 2.28

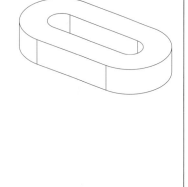

Editing Features in Parts (Feature Editing)

After you generate a part, all dimensions that were in the profile disappear. This part is called a base feature. You have three options when editing a feature's dimensions:

1. To edit a feature in the browser, double-click on the feature's name. The feature's dimensions appear. Select a dimension to change, and type a new value. Select another dimension, or press [←] to exit the command. The model updates.

2. Another way to edit a feature in the browser is to right-click on the feature's name and choose Edit from the pop-up menu that appears. The feature's dimensions appear. Select a dimension to change, and type a new value. Select another dimension, or press [←] to exit the command. Then update the model using the Update Part command (AMUPDATE), (the icon that resembles a lightning bolt) from the Part Modeling toolbar or from the bottom of the browser. Always update your part before adding or removing features; never save the part in an edited state, as this could corrupt the drawing.

3. Issue the Edit Feature (AMEDITFEAT) command from the Part Modeling toolbar (see Figure 2.29). Choose from the five options on the command line:

   ```
   Independent array instance, Sketch, surfCut, Toolbody, and
   <select Feature>:
   ```

Figure 2.29
Edit Feature command

Independent Array Instance

Independent array instance makes an array feature independent of the original array. This option works for polar and rectangular arrays. To use Independent array instance, issue the Edit Feature command (AMEDITFEAT) from the Part Modeling toolbar, then press I and press [←]. Next select the feature to edit. The feature you select has no link to the original array, and it cannot be rejoined to the original array unless you use the Undo command. However, you can edit or delete the independent feature. The feature is not constrained to the model. Use the Sketch option described in the following section to add necessary dimensions or constraints.

Viewing, Extruding, Revolving, Sweeping, and Editing Parts

Sketch

The Sketch option allows you to edit the original 2D sketch of a selected feature in addition to adding or deleting dimensions and constraints. To use Sketch, issue the Edit Feature command (AMEDITFEAT), press S, press [⏎], and select the part to edit. The part disappears and the 2D profile appears with the dimensions it had before it became a part. At this point you can add or delete constraints and dimensions, just as you would if it were a profile. After making the changes, issue the Update Part command (AMUPDATE) (the icon that resembles a lightning bolt) from the Part Modeling toolbar or from the bottom of the browser. The part regenerates, reflecting the changes to the sketch.

Notes: While editing a sketch, you can add and delete constraints and dimensions. The rules for working with 2D profiles apply in this edit mode.

To remove a dimension from a profile, use the Erase command.

Geometry cannot be added or deleted from a sketch in the Edit mode.

SurfCut

The third option in the Edit Feature command is surfCut, which allows you to edit a surface that was used as a cutting edge. A surfCut occurs when a nurbs surface is used to remove material from a part. To use surfCut, issue the Edit Feature command (AMEDITFEAT), press C, press [⏎], and select the feature in which a surfCut was performed. The surface that was used in the surfCut appears. You can then edit the surface by using grips, scale, stretch, or the Move command. After making the changes, issue the Update Part command (AMUPDATE). The part regenerates, reflecting your changes. The surfCut command is covered in Chapter 10.

Toolbody

The fourth option is to edit a feature that was used in the Combine command (AMCOMBINE). To use Toolbody, issue the Edit Feature command (AMEDITFEAT), press T, press [⏎], and select the part that another consumed; these parts are referred to as toolbodies. All dimensions appear on the part that was consumed in the Combine command (AMCOMBINE). Select the dimension to edit, type a new value, and press [⏎]. After making the changes, issue the Update Part command (AMUPDATE). The part regenerates, reflecting your changes. The Combine command is covered in Chapter 6.

Select Feature

The fifth and default option is to select a feature to edit. Issue the Edit Feature command (AMEDITFEAT) and select a feature to edit. Depending on the model and where you selected, you may be prompted at the command line to accept the highlighted feature or to press N and press [⏎] to cycle to the next feature. Continue cycling through the features until the correct feature is highlighted, then press [⏎]. All the dimensions that were used to constrain the feature appear, as well as a dimension for the extrusion distance, draft angle, or angle of revolution, depending on the selected feature.

If you select a hole feature, a dialog box appears, the same dialog box that was used to create the hole. If you created the hole from two edges, the Hole dialog box appears. After you make changes in the dialog box and select OK, two dimensions will appear. Select the dimension to edit and type a new value. Multiple features can be edited at once by selecting another feature instead of a dimension. To change a dimension, select the dimension, type a new value, and press [⏎]. Change as many dimensions as desired this way, then press [⏎] to exit the command. Issue the Update Part command (AMUPDATE) (see Figure 2.30). The part regenerates, reflecting the new values.

Figure 2.30
Update Part command

Notes: The location of the draft angle and the extrusion distance appear at the point selected on the feature.

The Update Part command (AMUPDATE) update command can also be issued from the bottom of the browser.

Tutorial 2.9—Using Edit Features

1. Start a new drawing.
2. Draw a rectangle.
3. Profile the rectangle.
4. Add dimensions to the rectangle as shown in Figure 2.31.
5. Change to an isometric view using the 8 key.
6. Issue the Extrude command (AMEXTRUDE).

Figure 2.31

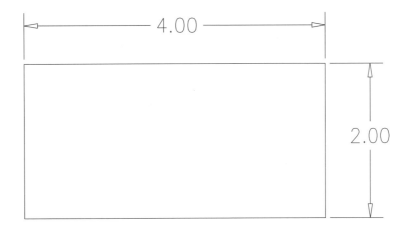

7. Extrude the rectangle 1 unit with 0 degrees of draft.
8. Accept the default direction for the extrusion.
9. Select the Edit Feature command (AMEDITFEAT).
10. Select the part.
11. Select the 4 inch length dimension, type "3", and press [Enter].
12. Select the 1 inch extrusion dimension, type "1.5", and press [Enter].
13. Select 0 for draft angle, type "-5", and press [Enter].
14. At the Select Object prompt, press [Enter] to exit the command.
15. Issue the Update Part command (AMUPDATE). Your drawing should resemble Figure 2.32.

Figure 2.32

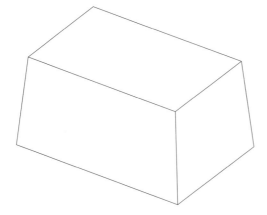

Mechanical Desktop 2.0: Applying Designer and Assembly Modules

16. Save your file as \Md2book\Drawings\Chapter2\EX2-9.dwg.

17. From the browser, double-click on the name ExtrusionBlind1, and change a few of the dimensions.

18. Press [↵] when done. The model should automatically update.

Tutorial 2.10—Editing a Feature

1. Continue working in drawing EX2-9, or open the file EX2-9.dwg.
2. Issue the Edit Feature command (AMEDITFEAT). The dialog box appears.
3. Type S, press [↵], and select the extrusion to edit the sketch.
4. Change to a plan view using the 9 key.
5. Issue the Delete Constraints command (AMDELCON). The dialog box appears.
6. Delete both vertical constraints by selecting both Vs. Your drawing should resemble Figure 2.33.

Figure 2.33

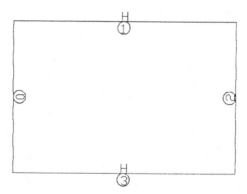

7. Press [↵] to exit the command. The constraints disappear and the dimensions reappear.

   Add two 70° dimensions as shown in Figure 2.34.

8. Change to an isometric view using the 8 key.
9. Update the model. Your drawing should resemble Figure 2.35.

Viewing, Extruding, Revolving, Sweeping, and Editing Parts

Figure 2.34

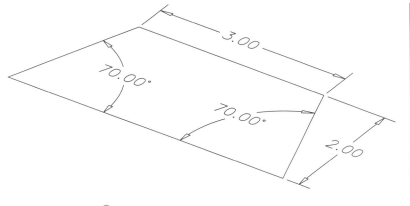

Figure 2.35

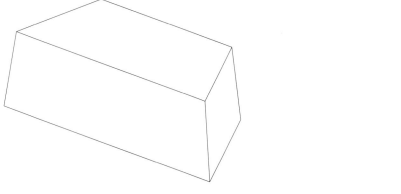

10. Save the file.

Exercises

The following section is intended to challenge you by providing open-ended problems. Like the problems you will encounter at work, each of the following problems has multiple solutions. To complete each exercise in this section, follow all steps, and save a file with the name of the drawing (e.g., \Md2book\Drawings\Chapter2\Bracket.dwg). Your solutions should match the accompanying figures.

The extrusion direction for all exercises is the positive Z direction.

Exercise 2.1—Bracket

1. Open the file \Md2book\Drawings\Chapter2\Bracket.dwg.

2. Extrude the bracket 3". Your drawing should resemble Figure 2.36.

Figure 2.36

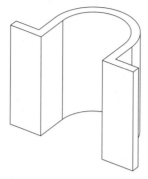

3. Save the file.

Exercise 2.2—Guide

1. Open the file \Md2book\Drawings\Chapter2\Guide.dwg.

2. Extrude it the profile 0.5".

3. Edit the feature so that the horizontal center-to-center distance is 4", as shown in Figure 2.37. (The 3" dimension is for reference only.) Remember to update the model after making the change.

Figure 2.37

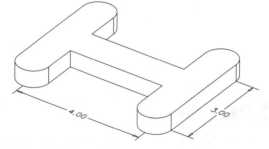

4. Save the file.

Exercise 2.3—Foot

1. Open the file \Md2book\Drawings\Chapter2\Foot.dwg.

2. Extrude the profile 3".

3. Edit the feature's "sketch" and delete the vertical constraint on the right line.

4. Add a dimension of 75°, as shown in Figure 2.38.

5. Update the model. Your drawing should resemble Figure 2.39.

Figure 2.38

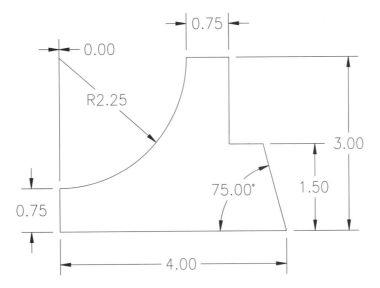

Figure 2.39

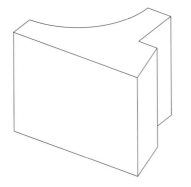

6. Save the file.

Exercise 2.4—Lamp

1. Open the file \Md2book\Drawings\Chapter2\Lamp.dwg.
2. Revolve the profile 360° about the inside vertical line.
3. Edit the feature's vertical dimensions, as shown in Figure 2.40.
4. Update the model. Your drawing should resemble Figure 2.41.
5. Save the file.

Figure 2.40

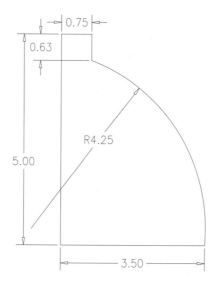

Figure 2.41

Exercise 2.5—Duct

1. Start a new drawing.

2. Create a path, as shown in Figure 2.42. The start point is the lower left line. Then dimension the path as shown in Figure 2.43.

3. Create a work plane with the Sweep option.

4. Sketch a square, and profile and dimension it, as shown in Figure 2.43.

5. Create a swept part and accept all defaults in the dialog box. Your model should resemble Figure 2.44.

6. Save the file as \Md2book\Drawings\Chapter2\Duct.dwg.

Viewing, Extruding, Revolving, Sweeping, and Editing Parts

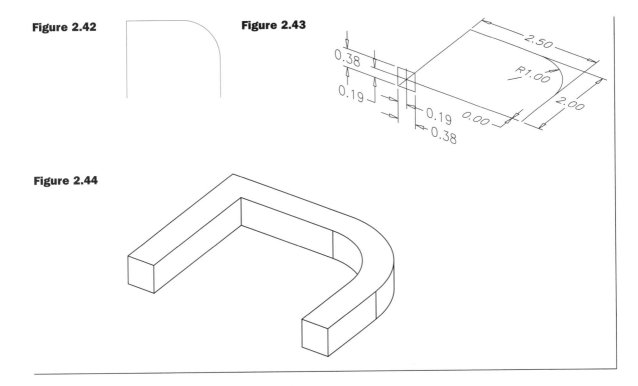

Figure 2.42

Figure 2.43

Figure 2.44

Review Questions

1. Name two methods for changing the viewpoint.
2. Describe the three methods for creating parts.
3. The 3 key changes the screen to three viewports with a top, a front, and an isometric view. T or F?
4. A model can only be edited in wireframe mode. T of F?
5. A sketch does not require profiling before extruding. T or F?
6. What is a base feature?
7. What objects can be used as an axis of revolution?
8. Differentiate between an extrusion and a sweep.
9. A profile does not need to be fully constrained. T or F?
10. Once a profile is extruded, you cannot delete or add constraints and dimensions to the profile. T or F?
11. You must update a part after a feature has been edited and before adding more features to the part. T or F?

chapter 3

Work Axis, Sketch Planes, Work Planes, Work Points and Visibility Options

Up to now, you have worked in a single plane in the world coordinate system. On that plane you have created your sketch, constraints, dimensions and then extruded, revolved or swept that profile into a part. The result was a base part, the first part created. When creating complex parts, you will create features that lie in different planes. In this chapter you will learn how to make other planes active and how to create planes that do not exist on the model.

After completing this chapter, you will be able to:

- Create a work axis.
- Make a plane the active sketch plane.
- Describe the differences between a sketch plane and a work plane.
- Create work planes.
- Create work points.
- Control the visibility of objects.

Creating a Work Axis

A work axis is a line that goes through the center of an arc or circle of a part. A work axis has three purposes:

1. It can be used to align the UCS of a selected sketch plane or work plane.
2. It is used to create a work plane that goes through the center of a cylindrical shaped part.
3. It is used as the axis of revolution in a polar array.

Figure 3.1
Work Axis command

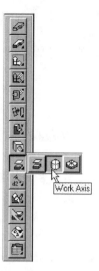

Before a work axis is created, a Mechanical Desktop part must exist and that part must contain an arc or circular edge. After these two conditions are satisfied, issue the Work Axis command (AMWORKAXIS) from the Part Modeling toolbar and select an arc or circular edge. A work axis will be created through the center of the selected geometry, extending beyond both sides of the part. If the arc or circular edge is present on both sides of the part, it does not matter which side of the geometry is selected. For example, to place a work axis in a cylinder, you could select either the top or bottom circular edge and you would get the same result. The work axis is tied to the arc or circular edge that was selected. If the location of the arc or circular edge changes, the work axis will also move.

Tutorial 3.1—Creating a Work Axis

1. Open the file \Md2book\Drawings\Chapter3\EX3-1.dwg.

2. Issue the Work Axis command (AMWORKAXIS).

3. Select the arcs and circle individually until they all have a work axis through them. Your screen should look like Figure 3.2.

4. Save the file.

Figure 3.2

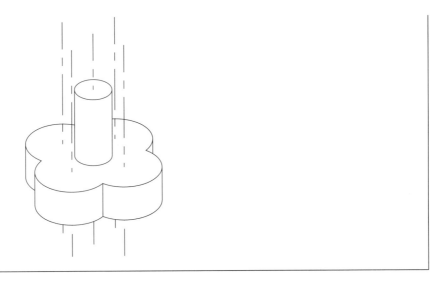

Making a Plane the Active Sketch Plane

A sketch plane is the active plane on which a profile is drawn. There can only be one active sketch plane at a time. Any face on a part or work plane can be made the active sketch plane. If you are familiar with the UCS command, you will see some similarities to placing the active sketch plane. However, do not use the UCS command, because Mechanical Desktop keeps track of the sketch plane in which the profile was drawn, and this is not the case with the UCS command.

There are two requirements in making a plane the active sketch plane: the part in which the plane will be placed must be a Mechanical Desktop part, and the part must contain a face or a work plane. The face does not need to have a straight edge. For example, a cylinder has two faces, one on the top and the other on the bottom of the part; neither has a straight edge. Issue the Create Sketch Plan command (AMSKPLN) from the Part Modeling toolbar. At the command line you will see options:

worldXy/worldYz/worldZx/Ucs/<Select work plane or planar face>:

The first four options refer to the World or current UCS. You will use these options only if there is no geometry to select from. The last option is for selecting a work plane or planar face. Work planes will be described later in this chapter. A planar face is a flat plane on your model. To make a plane active, select either an edge that helps define the plane or select anywhere inside the face. The selected plane will be highlighted. If the selected edge could belong to another plane or the selected point in the middle of the plane could refer to a plane behind the first plane, you be get prompted to cycle to the next plane. Cycle through the planes by pressing **N** and [⏎] until the correct plane is highlighted. Then press [⏎] to accept the highlighted plane. The next set of options will orient the UCS to the World X, Y, Z or to a selected edge or axis. At the command line you will see options;

`worldX/worldY/worldZ/<Select work axis or straight edge>:`

The first three options will be used only if there are no axes or edges available for alignment. The default is to select an axis, or an edge. After selecting an edge to align to you will be prompted:

`Rotate/Z-flip/<Accept>:`

Rotate: The Rotate option will rotate the X,Y coordinate in 90° increments.

Z-Flip: The Z-flip option flips the Z coordinate 180°.

Accept: Pressing [⏎] will accept the current orientation. To look straight down on current sketch view, also referred to as plan view, issue the Desktop View command (AMVIEW) or press the **9** key.

Figure 3.3
Create Sketch Plane command

Tutorial 3.2—Sketch Planes: Selecting in the Middle of a Plane

1. Start a new drawing.
2. Draw, profile, dimension and extrude a 2" cube.
3. Change to an isometric view with the 8 key.

4. Issue the Create Sketch Plan command (AMSKPLN).
5. Select in the middle of the front left plane and the plane should highlight as shown in Figure 3.4.

Figure 3.4

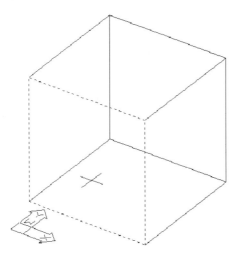

6. Press **N** and [⏎] to highlight the bottom or the back plane, depending on your selection point.
7. Press **N** and [⏎] to again highlight the front plane.
8. Press [⏎] to accept this plane.
9. To orient the UCS to this plane, select the bottom horizontal line of the front left plane of the cube, as shown with the square in Figure 3.5.

Figure 3.5

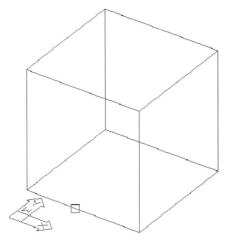

10. The UCS should be oriented as in Figure 3.6. If it is not, press **R** and [↵] to rotate the UCS until it does.

11. Save the file as \Md2book\Drawings\Chapter3\EX3-2.dwg.

Figure 3.6

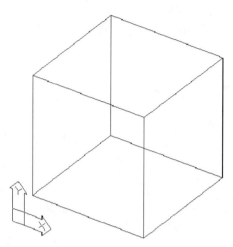

Tutorial 3.3—Sketch Planes: Edge Selecting

1. Continue with the same file as in Tutorial 3.2, or open the file \Md2book\Drawings\Chapter3\Ex3-2.dwg.

2. Issue the Create Sketch Plan command (AMSKPLN).

3. Select the lower edge of the front right plane as shown with the square in Figure 3.7.

4. Press **N** and [↵] to highlight the bottom plane.

5. Press **N** and [↵] to again highlight the front plane.

6. Press [↵] to accept this plane.

7. To orient the UCS to this plane, select the same edge as you did in step 3.

8. Press **Z** and [↵] to flip the Z direction 180°.

9. Press **Z** and [↵] to return to the original orientation.

Figure 3.7

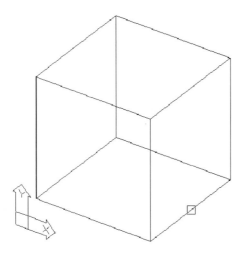

10. Press **R** and [Enter] four times to rotate the UCS 360° in 90° increments. When you are done, press [Enter] and the UCS should be back where it started and look like Figure 3.8.

11. Save the file.

Figure 3.8

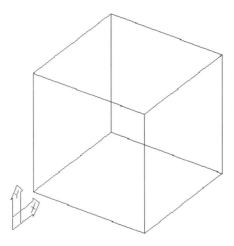

Mechanical Desktop 2.0: Applying Designer and Assembly Modules

Tutorial 3.4—Sketch Planes: Selecting with No Edges

1. Start a new drawing.

2. Draw a circle, profile it, give it a 2" diameter and extrude it 2".

3. Switch to an isometric view using the **8** key.

4. Issue the Create Sketch Plan command (AMSKPLN).

5. Select the top of the cylinder. Either select the top edge or in the middle of the face. You will not be prompted for the next plane because there is only one plane where you selected.

6. The next prompt is to orient the UCS. In this case, there is not an edge to select so you need to use the world coordinate options. Press **X** and ⏎.

7. Press **R** and ⏎ to rotate the UCS 90°. When complete, your drawing should look like Figure 3.9.

8. Save the file as \Md2book\Drawings\Chapter3\Ex3-4.dwg.

Figure 3.9

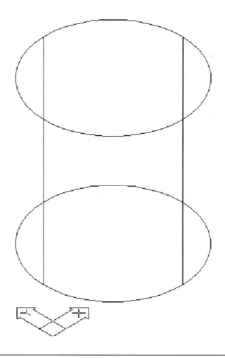

Creating a Work Plane

Before you consider work planes, it is important that you understand when you need to create a work plane. If there is a face on the model that can be used as the active sketch plane, make it the active sketch plane. If there is no face where you need one, then you create a work plane. A work plane can be made the active sketch plane and used as the plane for extruding, revolving or sweeping. A work plane is a rectangular plane that is geometrically tied to the part and that will always be larger than the part. If the part moves, the work plane will also move. For example, if a work plane is tangent to the outside face of a 1"-diameter cylinder and the cylinder diameter changes to 2", the work plane moves with the outside face of the cylinder. You can create as many work planes on a part as needed, and any work plane can be made the active sketch plane with the Create Sketch Plan command (AMSKPLN) from the Part Modeling toolbar.

A work plane is a feature and should be edited and deleted as a feature. For example, to edit an angle or offset distance of a work plane, use Edit Feature command (AMEDITFEAT) or double-click on the work plane's name in the browser. To delete work planes, use the Delete Feature command (AMDELFEAT) that will be covered in Chapter 4. Do not use the erase command, because this will delete the entire part.

Later in this chapter you will learn how to control the visibility of work planes. Before creating a work plane, ask yourself where this work plane needs to exist and what you know about this location. For example, you might want a plane to be tangent to a given face and parallel to another plane, or to go through the center of two arcs. Once you know what you want, select from the appropriate options and create a work plane. There are times when you may need to create an intermediate (construction) work plane before creating the final work plane. An example of this need: creating a work plane that is at 30° and tangent to a cylindrical face. First create a work plane that at a 30° angle and located at the center of the cylinder, and then create a work plane parallel to the angled work plane that is also tangent to the cylinder.

When work planes are created, it is best if they are created from a Mechanical Desktop part. They can also be created with the World Coordinate system. However, if you rotate the part in 3D space, you may get unexpected results with the work plane's not following the geometry.

To create a work plane, issue the Work Plane command (AMWORKPLN) and a dialog box will appear on the screen as shown in Figure 3.11. This dialog box consists of four sections: 1st Modifier, 2nd Modifier, Create Sketch Plane and an area for offset distance or angle. The 1st and 2nd Modifiers will determine how the work plane is tied to the part. As different options for the 1st Modifier are selected, there will be corresponding options that are grayed out, not available, in the 2nd Modifier area. It does not matter which modifier is selected for the first or second modifier, except for the Sweep Profile option, which does not require a second modifier. The Create Sketch Plane area can be

Mechanical Desktop 2.0: Applying Designer and Assembly Modules

checked if you want to make the new work plane the active sketch plane. After creating the work plane, you will be prompted to align the UCS to that plane. The last section is an area for specifying an offset distance or an angle.

Figure 3.10
Work Plane command

Figure 3.11
Work Plane dialog box

Notes: When aligning the UCS to the active sketch plane, select a work axis or an edge that is parallel to the sketch plane. If there is no axis or edge to select, refer to the World Coordinate system.

When the modifiers for creating a work plane are selected, it does not matter which option is the first or second modifier, except for Sweep Profile, which does not require a second modifier.

Work planes are tied to the part—if the part changes, the work plane will maintain the original modifiers used when it was created.

Work planes and work axis will always extend beyond the part; if the part gets larger, the work plane will re-size.

The best way get an understanding of work planes is to create them. In the tutorials that follow, you will create the most common types of work planes. Before each tutorial, there is a description of the type of work plane that will be created.

Work Planes On Edge/Axis - On Edge/Axis

In this exercise you will create two work axes and then create a work plane that passes through each axis. The options that you will use are On Edge/Axis and On Edge/Axis. An edge is any line in the part, and an axis refers to a work axis. With the modifiers On Edge/Axis and On Edge/Axis, you can create a work plane using these combinations: two edges, two work axes or an edge and a work axis.

Work Axis, Sketch Planes, Work Planes, Work Points and Visibility Options

Tutorial 3.5—Creating a Work Plane On Edge/Axis - On Edge/Axis

1. Open the file \Md2book\Drawings\Chapter3\Ex3-5.dwg.

2. Issue the Work Axis command (AMWORKAXIS) to create a work axis through each arc.

3. Issue the Work Plane command (AMWORKPLN) to create a work plane though each work axis. Use On Edge/Axis for both the 1st and 2nd Modifiers, check Create Sketch Plane and then select OK.

4. Select both work axes; it does not matter which axis is selected first.

5. To orient the UCS, select one of the axes and rotate the UCS until your screen looks like Figure 3.12.

Figure 3.12

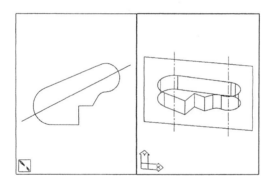

6. Press [⏎] to repeat the Work Plane command and create a work plane through the two front edges. Again set both the 1st and 2nd Modifiers to On Edge/Axis, check Create Sketch Plane and then select OK. The X's in Figure 3.13 show the edges to select; the order is not important.

7. To orient the UCS, select one of the edges and rotate the UCS until your screen looks like Figure 3.13.

8. Save the file.

Figure 3.13

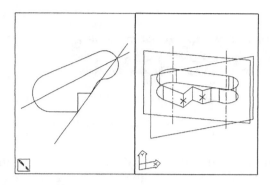

Work Planes On Edge/Axis - Planar Angle

In this tutorial you will create a work plane that is angled: you will select a work axis, or straight edge, that will act as the center of rotation and specify a plane as the reference for the angle. After the plane is selected, an arrow will appear showing the direction. You can either accept or flip the direction.

Tutorial 3.6—Creating a Work Plane On Edge/Axis - Planar Angle

1. Open the file \Md2book\Drawings\Chapter3\Ex3-6.dwg.

2. Issue the Work Axis command (AMWORKAXIS) and create a work axis through the left arc.

3. Issue the Work Plane command (AMWORKPLN) to create an angled work plane. Set the 1st Modifier to Work Axis and the 2nd Modifier to Planar Angle, type in "60" for the Angle, check Create Sketch Plane and then select OK.

4. Select the work axis.

5. Select the front plane as shown in Figure 3.14. When the front face is highlighted, press [Enter].

6. The work plane will appear on the screen. Flip the direction of the work plane by pressing **F** and [Enter].

7. Flip the direction by pressing **F** and press [Enter] twice to accept this direction.

8. To orient the UCS, select any vertical line and rotate it if necessary. When complete, your drawing should look like Figure 3.15. The edge you select for orienting the UCS does not have to be an axis or edge that you used to create the work plane.

Work Axis, Sketch Planes, Work Planes, Work Points and Visibility Options

Figure 3.14

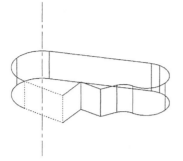

Figure 3.15

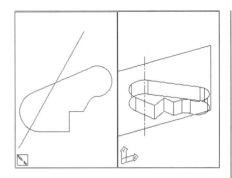

9. Change the angle of the work plane to 45° with the Edit Feature command (AMEDITFEAT): Issue the command, select the work plane or double-click on the name WorkPlane1 in the browser, select the 60° dimension and change it to 45. Then update the model with the Update Part command (AMUPDATE).

10. Save the file.

Work Planes: Tangent - Planar Parallel

In this tutorial you will create a work plane that is tangent to an arc and parallel to another work plane. To create a tangent work plane, you will select an arc or circular edge to be tangent to; it does not matter what side of the extrusion you select. Then you will select a plane to which you want the work plane to be parallel. After the work plane appears on the screen, you will prompted to either accept or flip the side to which the work plane will be tangent.

Tutorial 3.7—Creating a Work Plane: Tangent - Planar Parallel

1. Open the file \Md2book\Drawings\Chapter3\Ex3-7.dwg.

2. Issue the Work Plane command (AMWORKPLN) to create a Tangent work plane. Set the 1st Modifier to Tangent and the 2nd Modifier to Planar Parallel, check Create Sketch Plane and then select OK.

3. For the prompt Select cylindrical face:

    select the leftmost arc.

4. For the prompt Select work plane or planar face

    select the existing work plane.

5. Press **F** and [↵Enter] to flip the direction for the tangency and press [↵Enter] to accept this direction.

6. To orient the UCS, select any vertical line and rotate it if necessary. When complete, your drawing should look like Figure 3.16.

Figure 3.16

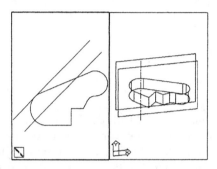

7. Press [↵Enter] to repeat the Work Plane command, set the 1st Modifier to Tangent and the 2nd Modifier to Planar Parallel, check Create Sketch Plane and select OK.

8. For the prompt: `Select cylindrical face:`

   select the rightmost arc.

9. For the prompt: `Select work plane or planar face:`

   select the face as marked with an X in Figure 3.17.

Figure 3.17

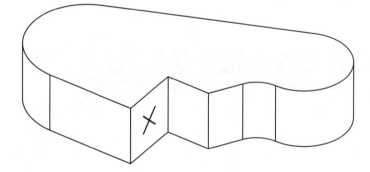

10. Press [↵Enter] to accept the default direction of the tangency.

11. To orient the UCS, select any vertical line and rotate it if necessary. When complete, your drawing should look like Figure 3.18.

12. Save the file.

Figure 3.18

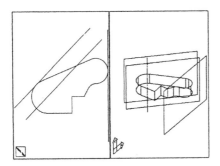

 Tutorial 3.8—Creating a Work Plane Tangent - Planar Parallel

1. Start a new drawing.
2. Create a 2"-diameter circle and extrude it 2".
3. Change to an isometric view using the **8** key.
4. Create a work plane with the 1st Modifier set to Tangent and the 2nd Modifier to Planar Parallel, check Create Sketch Plane and then click OK.
5. Select either the top or bottom circle (it does not matter where you select).
6. Since there is no axis or edge available as a parallel plane, you must refer back to the world coordinate system. In this tutorial, press **Y** and [↵] to orient it to the world YZ plane.
7. Press [↵] to accept the default direction.
8. To align the UCS, you must again refer back to the world coordinate system. In this exercise type "Z" to align it to the world Z axis. Rotate the UCS until it matches Figure 3.19.
9. Save the file as \Md2book\Drawings\Chapter3\Ex3-8.dwg.

Figure 3.19

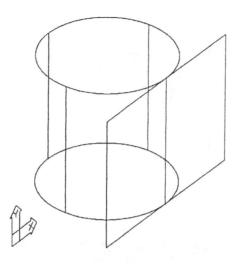

Work Planes: Tangent - On Edge/Axis

In this tutorial you will create a work plane that is tangent to an arc and attached to an edge and create another plane that is normal. To create a work plane that is tangent to an arc and tied to an edge, select on the side of the arc to which you want the work plane to be tangent.

Tutorial 3.9—Creating a Work Plane Tangent - On Edge/Axis

1. Open the file `\Md2book\Drawings\Chapter3\Ex3-9.dwg`.

2. Create a work plane with the 1st Modifier set to Tangent and the 2nd Modifier to On Edge/Axis, check Create Sketch Plane then select OK.

3. For the prompt `Select cylindrical face:`

   select near the bottom of the right arc, as shown in Figure 3.20.

4. For the prompt `<Select work axis or straight edge>:`

   select the vertical edge, as shown in Figure 3.20.

5. To orient the UCS, select any vertical line and rotate it if necessary. When complete, your drawing should look like Figure 3.21.

6. Save the file.

Work Axis, Sketch Planes, Work Planes, Work Points and Visibility Options

Figure 3.20

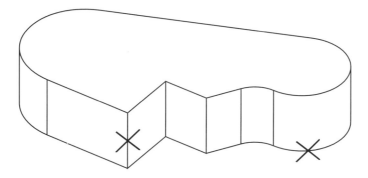

Figure 3.21

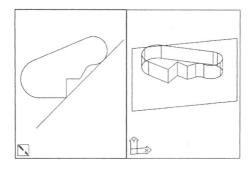

Work Planes: Planar Parallel

When creating a parallel work plane, you can create it from another work plane, a plane on the part, or parallel to the world coordinate system. The plane will be parallel to the selected plane and will be placed at the location specified by the second modifier.

Tutorial 3.10—Creating a Work Plane Planar Parallel - On Vertex

1. Open the file \Md2book\Drawings\Chapter3\Ex3-10.dwg.
2. Create a work plane with the 1st Modifier set to Planar Parallel and the 2nd Modifier set to Vertex, check Create Sketch Plane and then select OK.
3. Select the back face as highlighted in Figure 3.22.
4. Select the vertex as marked with an "X" in Figure 3.22.
5. To align the UCS, select any vertical line and rotate it if necessary. When complete, your drawing should look like Figure 3.23.

Figure 3.22

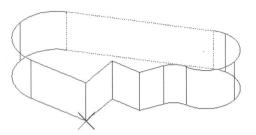

Figure 3.23

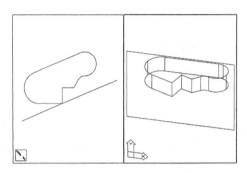

Tutorial 3.11—Creating a Work Plane Planar Parallel - Offset

1. Keep working with the file from Tutorial 3.10.

2. Create a Work Plane with the 1st Modifier set to Planar Parallel and the 2nd Modifier set to Offset, check the Create Sketch Plane, type in "2" for the Offset distance and then select OK.

3. Select the work plane that was created in Tutorial 3.10.

4. Press **F** and [←Enter] to flip the offset direction, and press [←Enter] to accept this direction.

5. To orient the UCS, select any vertical line and rotate it if necessary. When complete, your drawing should look like Figure 3.24.

6. Create a Work Plane with the 1st Modifier set to Planar Parallel and the 2nd Modifier set to Offset, check Create Sketch Plane, type in "1" for the Offset distance and then select OK.

7. Select the same face as marked in Figure 3.17.

Figure 3.24

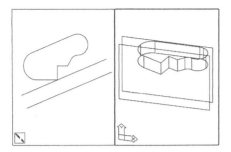

8. Press [⏎] to accept the default direction.
9. To orient the UCS, select any vertical line and rotate it if necessary. When complete, your drawing should look like Figure 3.25.

Figure 3.25

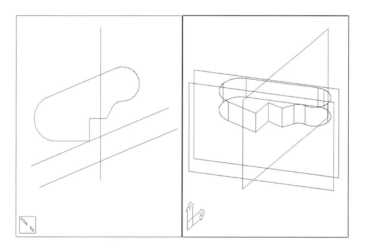

Work Planes: On Vertex

The On Vertex option is based on the principle that three points define a plane. In the next two tutorials, you will define planes based on this principle. Remember that an edge defines two vertices. When selecting a vertex, do not use an object snap; Mechanical Desktop will automatically find the closest vertex.

Tutorial 3.12—Creating a Work Plane On Vertex - On Edge/Axis

1. Open the file \Md2book\Drawings\Chapter3\Ex3-12.dwg.

2. Create a Work Plane with the 1st Modifier set to On Vertex and the 2nd Modifier set to On Edge/Axis, check Create Sketch Plane an then select OK.

3. Select the vertex of the back vertical line as marked with an "X" in Figure 3.26.

4. Select the edge of the lower horizontal line as marked with an "X" in Figure 3.26.

Figure 3.26

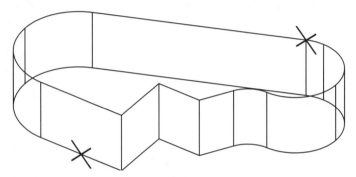

5. To orient the UCS, select the same edge as you did in step 3 and rotate the UCS if necessary. When complete, your drawing should look like Figure 3.27.

6. Save the file.

Figure 3.27

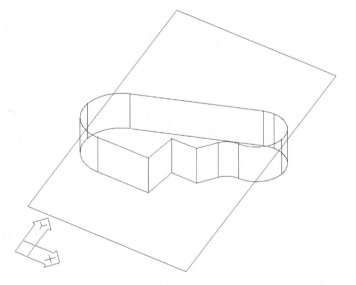

Tutorial 3.13—Creating a Work Plane On Vertex - On 3 Vertices

1. Keep working with the file from Tutorial 3.12.
2. Create a work plane with the 1st Modifier set to On Vertex and the 2nd Modifier set to On 3 Vertices, uncheck Create Sketch Plane and then select OK.
3. Select the vertices as shown in Figure 3.28. When selecting the vertices, the order is not important. Your drawing should look like Figure 3.29.

Figure 3.28

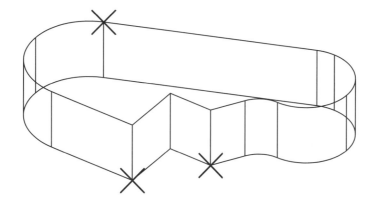

4. Save the file.

Figure 3.29

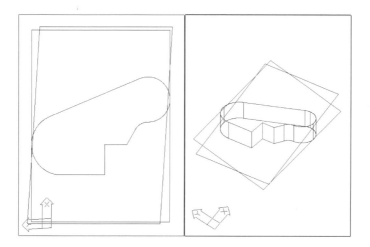

Creating Work Points

A work point is a feature consisting of three short lines, one in each axis (X, Y, and Z), that intersect at their midpoints. A work point can be used as the axis of rotation for polar arrays or for placing hole features. Arrays and holes will be discussed further in Chapter 4. To create a work point, issue the Work Point command (AMWORKPT) from the Part Modeling toolbar and select a point on the active sketch plane. Because the work point is a feature, it does not need to be profiled. The work point will not automatically be constrained; it will require two dimensions or constraints. When a work point is dimensioned, the dimension will go to the intersection of the three lines, no matter where you select on the work point. After the work point is dimensioned, the dimensions will appear on the screen until the work point is used for an array or hole. A concentric constraint can be applied to constrain a work point to the center of an arc or circle. To delete a work point, use the Delete Feature command (AMDELFEAT).

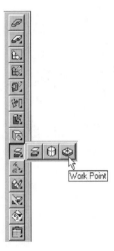

Figure 3.30
Work Point command

Tutorial 3.14—Creating Work Points

1. Open the file \Md2book\Drawings\Chapter3\Ex3-14.dwg

2. Issue the Work Point command (AMWORKPT) and place two work points as shown in Figure 3.31. The top plane is already the current sketch plane.

3. Apply a concentric constraint to the top circle and the rightmost work point.

4. Place two dimensions; ".75" and "0.0", as shown in Figure 3.31.

5. Make the work plane the active sketch plane with the Create Sketch Plan command (AMSKPLN) and orient the UCS around the world Z axis.

Work Axis, Sketch Planes, Work Planes, Work Points and Visibility Options

Figure 3.31

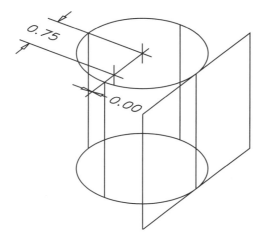

6. Issue the Work Point command (AMWORKPT) and create a work point on the work plane as shown in Figure 3.32.

7. Place two dimensions, "0.50" and "0.0", as shown in Figure 3.32. Both dimensions are dimensioned to the center of the circle.

8. Save the file.

Figure 3.32

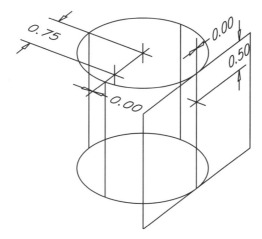

Controlling the Visibility of Objects

When working in AutoCAD, you can turn layers on and off to control the visibility of all objects on a particular layer. To control the visibility of object types or selected objects independently of the layer they are on, issue the Part Visibility command (AMVISIBLE). (Objects refer to work axes, work planes, work points, lines, arcs, circles etc.) After the command is issued, a dialog box appears, as shown in Figure 3.34. There are tabs along the top: the mode that you are working in (part, assembly, scene or drawing) will determine which tab will be current.

- Part modeling mode: where a part is created.
- Assembly modeling mode: where parts are assembled.
- Scene mode: where a scene is created for use with assemblies.
- Drawing mode: where drawing views are created.

Not all the tabs will be available at any one time. The scene tab is only visible when you are working in a scene.

To hide or unhide objects, select the Hide or Unhide button. The All box will hide/unhide all objects. To hide/unhide all objects except those selected, select the All button and the Except button. You will then be returned to AutoCAD to select objects. To hide an entire object group, check the box before the group. To hide specific object(s) select the Select option and you will be returned to AutoCAD to select the object(s). Select Apply to see the results without exiting the command or select OK to exit the command and see the results.

Under the Objects tab, you can control the visibility of AutoCAD objects like lines, arcs, circles, splines, parts, surfaces, etc. Another method for toggling the visibility of a work plane, work axis, work points and parts is to right click on its name in the browser and select Visibility from the pop-up menu. The browser will be covered in more depth in Chapter 4.

 Note: If a Mechanical Desktop drawing contains objects whose visibility has been turned off and it is opened in regular AutoCAD, those objects cannot be made visible through regular AutoCAD commands.

Figure 3.33
Part Visibility command

Work Axis, Sketch Planes, Work Planes, Work Points and Visibility Options

Figure 3.34
Visibility dialog box

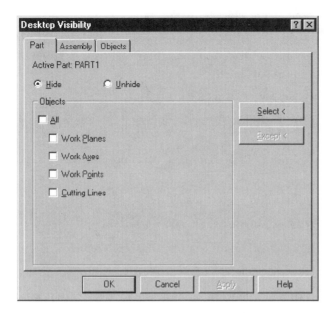

Tutorial 3.15—Using Visibility Options

1. Open the file EX3-15.dwg.
2. Issue the Part Visibility command (AMVISIBLE).
3. Hide all work planes and then select Apply.
4. Hide all work axes and then select Apply.
5. Unhide All and select Apply.
6. From the Object tab, hide all circles and then select Apply.
7. Hide one of the lines with the Select option.
8. Continue hiding and unhiding objects until you feel comfortable with the command.

Review Questions

1. A work axis can only be placed in the center of an arc or circular edge of a Mechanical Desktop part. T or F?
2. Name two purposes for creating a work axis.
3. Why should the Create Sketch Plane command be used instead of the UCS command?
4. How can you orient the UCS to a plane that does not have an edge defining the plane?
5. Explain a situation when a work plane needs to be created.
6. When creating a work plane, when will you not require a second modifier?
7. How do you make a work plane your active sketch plane?
8. After an angled work plane has been created, how can you change the angle?
9. In a drawing with multiple work planes, a single work plane cannot be turned off. T or F?
10. Visibility of AutoCAD objects can be controlled with Part Visibility command (AMVISIBLE). T or F?

chapter 4

Cut, Join and Intersect Operations, Browser Actions, Fillets, Chamfers, Holes and Arrays

In the last chapter, you learned about making a planar face or work plane your active sketch plane. In this chapter, you will create profiles on these planes and modify the part by adding to it or subtracting material from it. You will learn how to create placed features such as fillets, chamfers, holes and also how to array features.

After completing this chapter, you will be able to:

- Explain what is meant by the term "feature".
- Create and modify existing parts using one of the three operations: Cut, Join and Intersect.
- Use the browser to edit, rename, copy, delete and reorder features.
- Create four types of fillets.
- Create chamfers.
- Create holes.
- Create and edit both polar and rectangular arrays of features.

What Is a Feature?

In the previous chapters, you created base features, work planes, work axis and work points, all of which are features. There are also sketched features, where you draw a sketch on a planar face or work plane and either add material to the part or subtract material from the part. Extruding, revolving or sweeping can create these sketched features. You can also create placed features such as fillets, chamfers, and holes. Placed features require a base part to exist; they are added to the part. As the features are added to the part, they will appear in the browser, showing the history of the part (the order in which the features were created). Features can be edited or deleted from the part as required.

Cut, Join and Intersect Operations

Up to now, the parts you created have not been complex in shape; the base part is usually a simple extrude, revolve or sweep. To create more complex parts, you can create sketch features and do one of three operations: cut, join or intersect.

Cut: Removes material from the part.

Join: Adds material to the part.

Intersect: Keeps what is common to the part and this second feature.

Extruding, revolving or sweeping can be used on sketched features. Figure 4.1 shows the three operations. The top row represents a part on the left and a profile on the right. The bottom row shows the resulting material after this new feature is created.

Figure 4.1

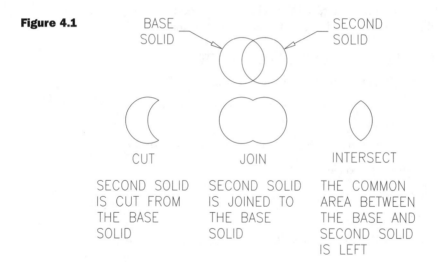

There are seven basic steps that you will follow when creating parts with sketched features.

1. Create a base part.
2. Create work axis and work planes as needed.
3. Make a planar face or work plane your active sketch plane.
4. Create a sketch of the geometry to be profiled.

Cut, Join and Intersect Operations, Browser Actions, Fillets, Chamfers, Holes and Arrays

5. Profile the sketch.

6. Constrain and dimension the profile.

7. Extrude, revolve or sweep the profile, removing from the part, adding to the part, or keeping what is common to the base part. Continue with steps 2-7 until the part is complete.

There are no limits to the number of features that can be added to a part. When dimensioning and constraining a profile, it is sometimes easier to work in a plan view (looking straight at the current plane) of the current sketch plane. The **9** key is programmed to do this, as well as the Desktop View command (AMVIEW). You can apply dimensions and constraints exactly as you did with the first profile, with the addition of constraining and dimensioning the profile to the existing part. You can also create dimensions by selecting geometry that does not lie on the current sketch plane, even though the dimensions will be placed on the current sketch plane. When you look at a part from different viewpoints, you will see arcs and circular edges appearing as lines. Remember that they are still circular edges; when you constrain or dimension to them, the constraints and dimensions will go to their center points. After constraining and dimensioning the profile, extrude, revolve or sweep the profile.

Before you extrude, revolve or sweep a profile, a dialog box will appear when the command is issued; you will need to fill in information for the Operation type and Termination type. Descriptions follow for both Operation and Termination that will be needed for extruding, revolving or sweeping.

Operation: Here you select the type of operation you need. There are four options to choose from. The Base option is the default for the first part and will be grayed out after it has been created.

Option:	Function:
Base:	The first feature. After a part is created from a profile, this option will be grayed out.
Cut:	Material will be removed.
Join:	Material will be added.
Intersect:	The material that is common to the part and the new feature will remain.

Termination: Here you specify the boundary of the feature being created. There are six options to choose from.

Option:	Function:
Blind	Extrudes a specific distance in the positive or negative Z direction. (Extrude only)
Through	Goes all the way through the part in one direction. (Extrude only)
To Plane	Continues until the profile reaches a specific plane face or work plane. (Extrude, Revolve only)
To Face	Continues until the profile reaches a specific face that is contoured.
From To	Starts at one plane/face and stops at another.
Mid Plane	Goes equal distances in the negative and positive directions. For example, if the extrusion distance is 2", the extrusion goes 1" in the negative and positive Z directions. (Extrude, Revolve only)
By Angle	Revolves the profile a specific number of degrees in the positive or negative Z direction. A prompt directs you to choose the direction in which to revolve. (Revolve only)
Full	Revolves the profile 360°. (Revolve only)
Path Only	Causes the profile to follow the path from its start point to its end point. (Sweep only)

Notes: There are no limits to the number of features that a part can have.

All features are displayed in the browser in the order in which they are created.

A collinear constraint can be applied between a line and an arc or circular edge of a feature.

Tip: When creating complex parts, think about how the finished part will look and break it down into simple shapes. Then create the part one shape at a time, adding and removing material as you go. If you are new to part creating, start with simple parts and work up to more complex parts.

Cut, Join and Intersect Operations, Browser Actions, Fillets, Chamfers, Holes and Arrays

To edit any feature, use the Edit Feature command (AMEDITFEAT) or the Desktop browser, which will be covered in more detail later in this chapter.

Each of the following tutorials guide you through a variety of methods for creating features. Remember, there is no one way to generate a specific part. After completing the tutorial as shown, go back and try your own ideas. Think about what the finished part will look like and create that part one feature at a time.

Tutorial 4.1—Extruding with Cut: Through

1. Open the file \Md2book\\Drawings\Chapter4\Ex4-1.dwg.
2. Make the front face the active sketch plane and orient the UCS as shown in Figure 4.2.

Figure 4.2

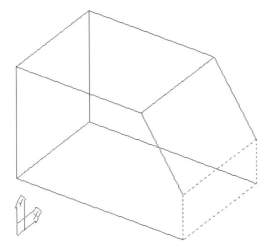

3. Change your view to the current sketch view (the **9** key).
4. Draw a rectangle and then profile, constrain (Collinear, top lines of the part and profile) and dimension the rectangle as shown in Figure 4.3.
5. Switch to an isometric view (the **8** key).
6. Issue the Extrude command (AMEXTRUDE) and use the following settings:

   Operation = Cut

Figure 4.3

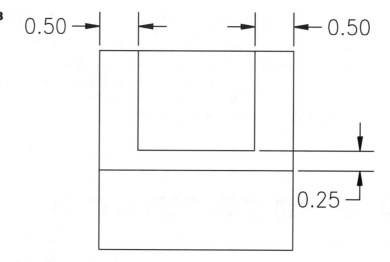

Termination to = Through

Draft Angle to = 0

Then select OK.

7. Press [⏎] to accept the default direction into your part. Your part should look like Figure 4.4, shown with lines hidden.

Figure 4.4

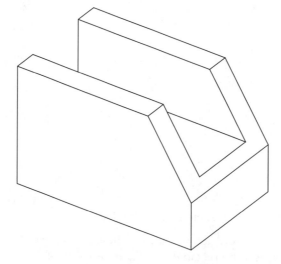

8. Edit the extrusion base length from "4" to "6" and update the part. The cut goes through the part even if the part's dimensions change.

9. Save the file.

Cut, Join and Intersect Operations, Browser Actions, Fillets, Chamfers, Holes and Arrays

Tutorial 4.2—Extruding with Cut: Blind

1. Open the file \Md2book\Drawings\Chapter4\Ex4-2.dwg.
2. Make the large outside circle (shown with the pickbox on it in Figure 4.5) the active sketch plane and orient the UCS (use the World Y option). When done, your part should look like Figure 4.5.

Figure 4.5

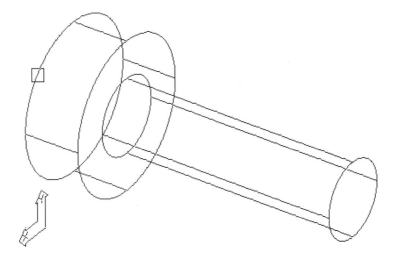

3. Change your view to the current sketch view (**9**).
4. Draw a rectangle and then profile, constrain (vertical lines tangent to the sides of the circle) and dimension the rectangle as shown in Figure 4.6.

Figure 4.6

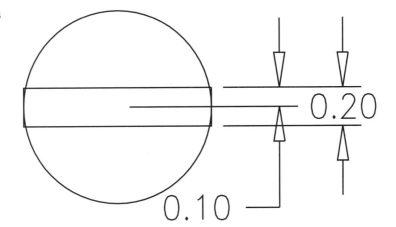

5. Switch to a southwest isometric view (**88**).

6. Issue the Extrude command (AMEXTRUDE) and use the following settings:

   Operation = Cut

   Termination = Blind

   Distance = .25

   Draft Angle = 0

   Select OK.

7. Press [⏎] to accept the default direction into your part (going into the part). Your part should look like Figure 4.7, shown with lines hidden.

Figure 4.7

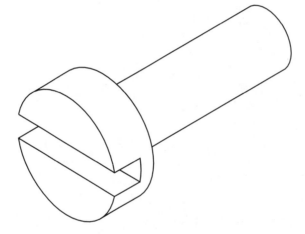

8. Save the file.

Tutorial 4.3—Revolving with Cut: Full

This file already has a work axis, work plane and a sketch that has been profiled.

1. Open the file \Md2book\Drawings\Chapter4\Ex4-3.dwg.

2. Change your view to the current sketch view (**9**).

3. For clarity, you may want to turn off the work plane. Work planes do not need to be visible, even if the current sketch plane is dependent on the work plane.

4. Place the dimensions as shown in Figure 4.8.

Figure 4.8

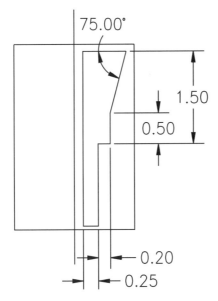

5. Switch to an isometric view (**8**).

6. Place a collinear constraint to the inside vertical edge of the profile and the work axis.

7. Place a collinear constraint to the top horizontal line of the profile and the top of the cylinder and to the bottom horizontal line of the profile and the bottom of the cylinder. This will fully constrain your part. A collinear constraint can be applied between a line and an arc or circular edge of a feature.

8. Issue the Revolve command (AMREVOLVE) and use the following settings:

   Operation = Cut

   Termination = Full

   Then select OK.

9. For the revolution axis, select either the work axis or the inside vertical line. Your part should resemble Figure 4.9.

10. Edit the extrusion distance of the cylinder to 3.5 and update the part. The revolved feature still goes to the end of the cylinder because the collinear constraint means that the two edges will always be in the same line, no matter what changes occur to the part.

11. Save the file.

Figure 4.9

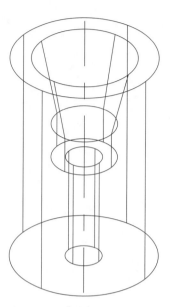

Tutorial 4.4—Sweeping with Cut: Path Only

This file already has a part, with an offset work plane in which a path was drawn, profiled, and dimensioned. A sweep work plane was then created and a circle was drawn, profiled and dimensioned.

1. Open the file \Md2book\Drawings\Chapter4\Ex4-4.dwg.
2. For clarity you may want to turn off the work planes.
3. Issue the Sweep command (AMSWEEP) and use the following settings:

   Operation = Cut

   Body Type = Normal

   Termination = Path Only

   Draft Angle = 0

   Then select OK.

4. When complete, your part should look like Figure 4.10.
5. Save the file.

Figure 4.10

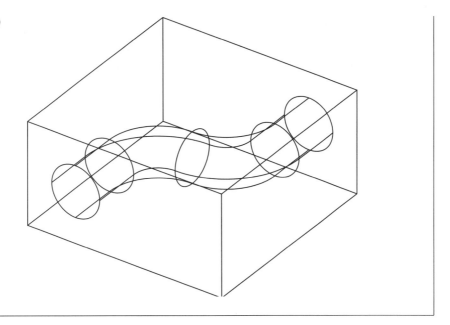

Tutorial 4.5—Extruding with Join: Blind

1. Open the file \Md2book\Drawings\Chapter4\Ex4-5.dwg.
2. Make the right vertical face the active sketch plane and orient the UCS, as shown in Figure 4.11.

Figure 4.11

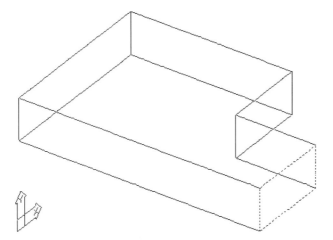

3. Change your view to the current sketch view (**9**).

4. Draw the tombstone shape and then profile, constrain (lines collinear to the sides of the extrusion) and dimension it as shown in Figure 4.12.

Figure 4.12

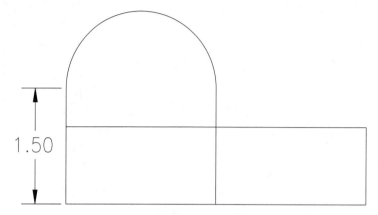

5. Switch to an isometric view (**8**).

6. Issue the Extrude command (AMEXTRUDE) and use the following settings:

   Operation = Join

   Termination = Blind

   Distance = 4

   Draft Angle = 0

   Then select OK.

7. Press F to flip the extrusion direction into the part and then press [⏎] to accept this direction. When complete, your part should look like Figure 4.12, shown with lines hidden.

8. Save the file.

Figure 4.13

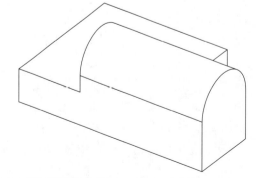

Tutorial 4.6—Extruding with Join: To Plane

1. If the file \Md2book\Drawings\Chapter4\Ex4-5.dwg is not the current file, open it now.

2. Make the bottom face the active sketch plane and orient the UCS, as shown in Figure 4.14. You may have to do a Z flip to orient the UCS in the positive direction.

Figure 4.14

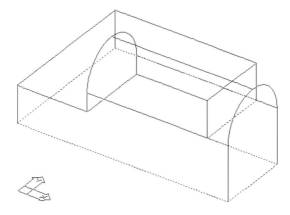

3. Switch to a southwest isometric view (**88**).

4. Draw the tombstone shape and then, profile, constrain the back horizontal line (collinear with the middle horizontal line of the extrusion) and dimension it as shown in Figure 4.15. The "0" dimension aligns the center of the arc to the line. The project constraint could have been used instead of the "0" dimension.

Figure 4.15

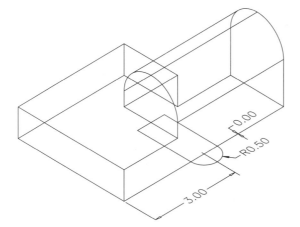

5. Create a Work Plane with the 1st Modifier set to Tangent and the 2nd Modifier to Planar Parallel, **uncheck** Create Sketch Plane and then select OK.

6. Select one of the extruded arcs for the cylindrical face.

7. Select the bottom plane for the planar face.

8. Press **F** to flip the direction toward the top of the part and then press ⏎ to accept this direction. Your part should look like Figure 4.16.

Figure 4.16

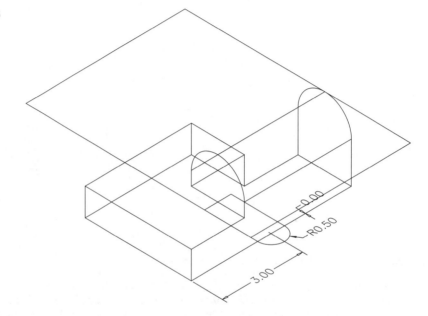

9. Issue the Extrude command (AMEXTRUDE) and use the following settings:

    Operation = Join

    Termination = Plane

    Draft Angle = 0

    Then select OK.

10. Select the work plane that you just created. Your part should look like Figure 4.17.

11. Edit the length of the first tombstone from "1.5" to "3" and update the part. The second extrusion height will always be tangent to the top of the first tombstone, because the work plane it is extruded to is tangent to that arc.

12. Save the file.

Figure 4.17

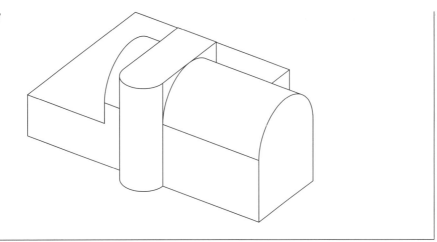

 Tutorial 4.7—Revolving with Join: From To

In this exercise you will use From To as the Termination option. This allows you to start and stop the extrusion, revolve or sweep at two different faces. Faces do not have to be planar, they can be contoured. In this tutorial, the profile (in green) has already been drawn and constrained. There is also construction geometry (the red centerline) around which the part will revolve. Construction geometry will be covered in Chapter 5. When performing a From To termination, zooming in closely will help you isolate graphically the face that is selected. You will have the option to cycle among the possible selection sets. You will select the first face and press [↵Enter] and then select the second face and press [↵Enter] to finish the command.

1. Open the file \Md2book\Drawings\Chapter4\Ex4-7.dwg.

2. Issue the Revolve command (AMREVOLVE) AND USE THE FOLLOWING SETTINGS:

    Operation = Join

    Termination = From To

    Draft Angle = 0

    Then select OK.

6. Select the red centerline as the axis of revolution.

7. Select the right inside vertical plane as shown in Figure 4.18 (highlighted plane). You may need to use the Next option to highlight the correct plane and then press [↵Enter].

Figure 4.18

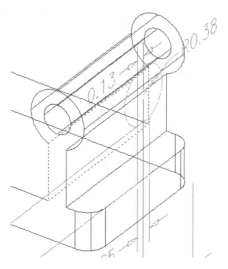

8. Select the left inside vertical plane as shown in Figure 4.19. You may need to use the Next option to highlight the correct plane and then press [Enter]. Your part should look like Figure 4.20.

Figure 4.19

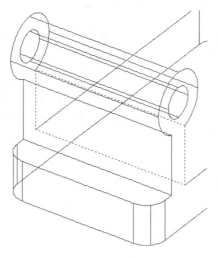

9. Save the file.

Rotate your part in shaded mode to verify that the revolved profile does not go through the hole. It should have stopped at the outside faces. If it does not look correct, redo the exercise, paying close attention to the faces that are selected for the start and ending termination faces.

Figure 4.20

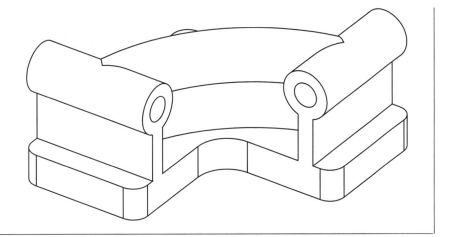

 Tutorial 4.8—Extruding with Intersect: Through

The sketch has already been profiled and dimensioned.

1. Open the file \Md2book\Drawings\Chapter4\Ex4-8.dwg.
2. Issue the Extrude command (AMEXTRUDE) and use the following settings:

    Operation = Intersect

    Termination = Through

    Draft Angle = 0

    Then select OK.

3. Press **F** and Press ⏎ to Flip the extrusion distance, then press ⏎ to accept this direction. When complete, your part should look like Figure 4.21.
4. Save the file.

Figure 4.21

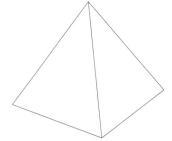

Tutorial 4.9—Revolving with Intersect: Full

The sketch has already been profiled and dimensioned.

1. Open the file \Md2book\Drawings\Chapter4\Ex4-9.dwg.
2. Issue the Revolve command (AMREVOLVE) and use the following settings:

   Operation = Intersect

   Termination = Full

   Then select OK.
3. For the revolution axis, select the green vertical line of the profile. When complete, your part should look like Figure 4.22.

Figure 4.22

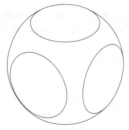

4. Issue the Edit Feature command (AMEDITFEAT) and edit the revolution: change the radius from "1.25" to "1.5" and change the distance the arc is offset from the cube from ".25" to ".5" on both sides. Update the part; it should look like Figure 4.23.
5. Save the file.

Figure 4.23

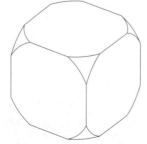

Using the Browser for Creating and Editing

In the first few chapters, you were briefly introduced to the browser. In this section, you will learn how to make use of the power of the browser when creating and editing parts. By default, the browser is docked along the left side of the screen. The browser itself acts like a toolbar, except that it can be resized while docked. If the browser is not showing, you can display it by issuing the Desktop Browser command (AMBROWSER) or by choosing Desktop Browser from the View pull-down menu. There are three possible tabs along the top of the browser; Assembly, Scene and Drawing. In Mechanical Desktop 2.0, when you start a new drawing you are working in an assembly, even if you only want to create one part. This can be confusing: even though you are working with a single part, you will be using the information under the Assembly tab of the browser. The majority of your time will be spent under the Assembly tab. If you want to work on a single part and not add parts to the file, you can start a new file with the New Part File option from the File pull-down menu. In that case, the assembly tab will be replaced with a part tab. In that mode you can only have a single part.

As you build a part, the history of the part will appear under the Assembly or Part tab. Here you can edit, rename, copy, delete and reorder features and parts. You can expand or collapse the history of the part(s) (the order in which the features and parts were created) by selecting the + and - on the left side of the part name in the Desktop browser. As parts grow in complexity, so will the information found in the browser. Dependent features will be indented to show that they are related to the top level. This is referred to as a parent-child relationship. The child cannot exist without the parent and is dependent on the parent. For example, if a hole is created in an extruded rectangle and the extrusion is then deleted, the hole will also be deleted. Each feature is given a default name. For example, a blind extrusion would be named ExtrusionBlind1. The browser can also help you locate parts and features in the drawing area. Select the feature or part name in the browser with your left mouse button and it will be highlighted in the drawing area.

Below you will find a description of the functionality found under the assembly or part tab.

Editing Features in the Browser Assembly or Part Tab

To edit a feature there are two methods. For the first method, double-click on the feature's name and, depending on the feature that is being edited, you will see either the dimensions or the dialog box in which it was created. After you edit the feature, the part will automatically be updated. The second method is to right-click on the feature name in the browser and choose Edit from the pop-up menu. After you edit the feature, the part will not automatically be updated. Select the Update Part command (AMUPDATE) from either the Part Modeling toolbar or on the bottom of the browser.

Mechanical Desktop 2.0: Applying Designer and Assembly Modules

Notes: When you start a new file with the NEW command, you can have one or more parts in the same file.

When you start a new file with the New Part File option from the File pull-down menu, you can only have a single part in that file.

Tip: To find a feature in a part, select the feature's name in the browser and it will be highlighted in the drawing.

Renaming with the Browser Assembly Tab

As previously explained, each feature and part is given a default name. You can set part name prefixes using the Edit Preferences command (AMPREFS). You can also change the name after it has been created in two ways. For the first method, slowly click twice on the feature name but, instead of letting go on the second click, keep it depressed for a second longer. The name will get shaded blue and a box will appear around the name. Rename the feature as you would using the Microsoft Explorer. If the double click is too fast, you will edit the feature. The second method is to right-click on the feature or part name and choose rename from the pop-up menu. Renaming a part through the browser does not change the part definition name only the name that the browser uses. To change the name of a definition, use the Desktop catalog. The Desktop catalog will be covered in Chapter 7.

Tutorial 4.10—Editing and Renaming through the Browser

1. Open the file \Md2book\Drawings\Chapter4\Ex4-10.dwg.
2. Select the + sign on the left side of the name EX4-10.
3. Select the + sign on the left side of the name Part1_1.
4. Select the + sign on the left side of the name ExtrusionBlind1.
5. Select the + sign on the left side of the name ExtrusionThru1.

   When you finish with step 5, your drawing should look like Figure 4.24.

6. To edit the blind extrusion, double-click on the name ExtrusionBlind1.
7. Change the extrusion depth from 1 to 2, and press [⏎] twice. The part will automatically be updated and should look like Figure 4.25.

Figure 4.24

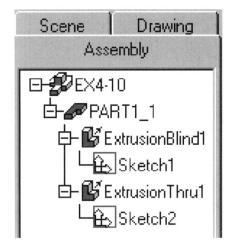

Figure 4.25

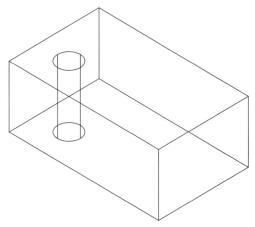

8. To edit the through extrusion, right-click on the name ExtrusionThru1 and choose Edit from the pop-up menu.

9. Change the diameter from .75 to 2 and the X and Y dimension from 1 to 1.5 and press [⏎] twice when done.

10. Update the part by selecting the Update Part (AMUPDATE) icon on the bottom of the browser.

11. Rename Part1 to Plate1 by slowly selecting the name twice. When the name is highlighted, type "Plate1".

12. Rename ExtrusionThru1 to Hole1 by right-clicking on the name and choosing Rename from the pop-up menu. When the name is highlighted, type "Hole1".

13. Save the file.

Figure 4.26

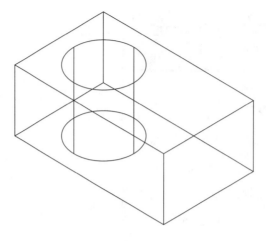

Copying with the Browser Assembly Tab

Once a feature has been created, it can be copied through the browser. Before copying a feature, make the plane where the copy of the feature will be placed the active sketch plane. The features can only be copied to the active part. The base part cannot be copied as a feature. To copy a feature, select its name from the browser with a right mouse click, then choose the Copy option. You will be returned to the drawing, where you are prompted to select a location on the current sketch plane. Before accepting a location, you have options to rotate the feature around the Z axis, flip the feature around the XY plane, or press [⏎] to accept this location. After the feature is copied, it is not yet constrained to the part. You can accomplish this by editing the sketch of the feature and adding dimensions and constraints; this will be covered in Chapter 5.

Deleting with the Browser Assembly Tab

To erase a feature, select its name with the right mouse button and choose Delete from the pop-up menu. At the command line you will be prompted:

```
Highlighted features will be deleted. Continue ? No/<Yes>:
```

Press [⏎] to delete the feature or press **N** to return to the command line without deleting the feature. If you delete a parent feature with a child feature, you will be alerted to the fact that the child will also be deleted. You will have the option to cancel or continue the operation.

Cut, Join and Intersect Operations, Browser Actions, Fillets, Chamfers, Holes and Arrays

Tutorial 4.1—Copying and Deleting through the Browser

1. Open the file \Md2book\Drawings\Chapter4\Ex4-11.dwg.
2. Expand all sections in the browser by selecting all the + signs.
3. Activate the Part Modeling toolbar if it is not active.
4. Make the front face the active sketch plane and align the UCS with the positive X direction going into the screen, see Figure 4.27.

Figure 4.27

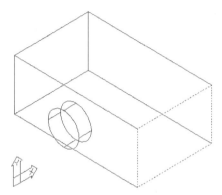

5. Select ExtrusionBlind2 with the right mouse button and choose the Copy option from the pop-up menu.
6. Select a point near the middle of the front face as shown in Figure 4.28 and press ⏎.

Figure 4.28

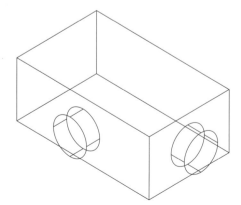

7. To delete the first extrusion, right-click on the name ExtrusionBlind2 and choose the Delete option from the pop-up menu.

8. Press [enter] to accept this selection. Your drawing should look like Figure 4.29.

9. Save the file.

Figure 4.29

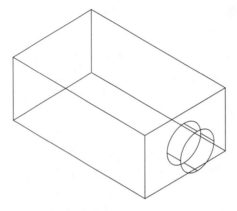

Reordering through the Browser

When generating a complex part, you may at some point create a feature and not get the expected result. After analyzing the part, you may determine that if the feature were created before or after another feature, you would have achieved the expected result. You do not have to recreate the part. Instead, you will change the position of the feature in the browser. This is referred to as reordering a feature. To reorder a feature, select its name in the browser with the left mouse button and, keeping the button depressed, drag it up or down until it is in the correct location. As you drag the name in the browser, you will see a circle with a diagonal line through it. This means that the location is not a valid one. Keep moving the mouse until a horizontal line is displayed in the browser, showing this is a valid location. Then release the mouse button and the part will update with its new order. There is more information on reordering in Chapter 6.

Cut, Join and Intersect Operations, Browser Actions, Fillets, Chamfers, Holes and Arrays

Tutorial 4.12—Reordering through the Browser

In this tutorial, a rectangle was extruded, a through hole was created and then the part was shelled out. The hole became a boss from the shell, as shown in Figure 4.30. You will reorder the hole so that its order is after the shell, again making it a through hole. The shell command will be covered in Chapter 6.

Figure 4.30

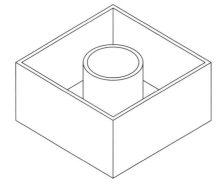

1. Open the file `\Md2book\Drawings\Chapter4\Ex4-12.dwg`.

2. Expand all sections in the browser.

3. From inside the browser, hold the left mouse button down while it is over the name Hole1. Drag the name Hole1 below the name Shell1, as shown in Figure 4.31. As you move the mouse, you will see a circle with a diagonal line through it, telling you that it can not be placed in that location. Keep moving it until a horizontal line is displayed across the browser, showing that this is a valid location.

4. Save the file.

Figure 4.31

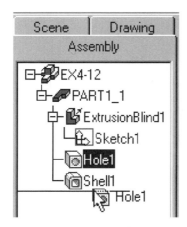

Figure 4.32

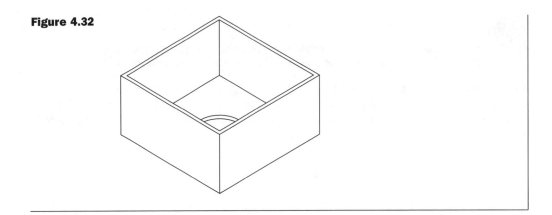

Fillets

When creating fillets in 3D, you will select the edge that needs to be filleted and the fillet will be created between the two faces sharing this edge. This differs from filleting in 2D, because in 2D you select two objects and a fillet is created between them. When creating a part, it is usually good practice to create fillets and chamfers as one of last features in the part. Fillets add complexity to the part, which in turn adds to the file size and removes edges that may be needed to place other features. After the Fillet command (AMFILLET) is issued, a dialog box appears. Select from four types of fillets (constant, fixed width, linear or cubic) and then type in a size if it is a constant or fixed width fillet. The four types of fillets will be described in the next section. To edit a fillet you can use either the Edit Feature command (AMEDITFEAT) or the browser, as described earlier in this chapter. Select the fillet to edit and the value of its radius will appear. Select the radius that you want to change and then type in a new value and press [↵]. Keep selecting the different radiis that you want to change, pressing [↵] to exit the command, and then update the part.

Figure 4.33

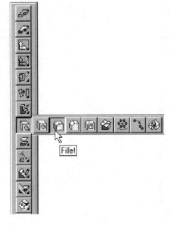

Constant Fillet

A constant fillet has the same radius from the beginning to the end of the fillet. Issue the Desktop Fillet command (AMFILLET) from the Part Modeling toolbar. From the dialog box select Constant, type in a value for the radius, select Apply and then select the edge(s) that you want to fillet. There is no limit to the number of edges that can be filleted with a constant fillet. The order in which the edges are selected is not important. The edges that are to be filleted need to be picked individually, the use of window, crossing, etc. is not valid for creating fillets. When you are done selecting the edge(s) to fillet, press [←Enter]. If you select multiple edges in the same command, they are linked together. If you edit a fillet that was selected as a group, they will all get highlighted. Change the value and they all change. Delete one fillet of the group and they all get deleted. If you think you would like to edit them independently of the group, check the box Individual Radii Override under Constant and then each fillet can have a different radius. In the Desktop browser, the fillets are grouped under one name, but when you edit them you can change their values independently.

At the bottom of the dialog box is a section Return to dialog box. If you want to create multiple fillets one after another, check here and you will be returned to the dialog box after you create the fillet.

Tutorial 4.13—Creating a Constant Fillet

1. Open the file `\Md2book\Drawings\Chapter4\Ex4-13.dwg`.
2. Issue the Desktop Fillet command (AMFILLET).
3. Check Constant, Return to dialog box, type in a radius of .5, if it is not already .5, and select Apply.
4. Select the front vertical edge and press [←Enter].
5. Back in the dialog box, change the radius to .25, **uncheck** Return to dialog box and select Apply.
6. Select all the remaining edges and press [←Enter]. When complete, your drawing should look like Figure 4.34.
7. Edit the ".5" fillet to ".25" using the Edit Feature command (AMEDITFEAT).
8. Edit the ".25" fillets to ".125" by double-clicking on the name Fillet2 in the browser.
9. Update the part. It should resemble Figure 4.35.
10. Delete the group of ".125" fillets, update the part and your part should resemble Figure 4.36.

Figure 4.34

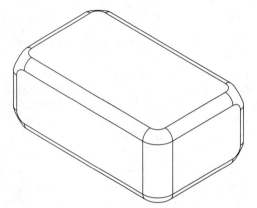

Figure 4.35

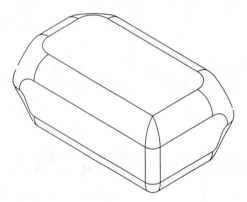

Figure 4.36

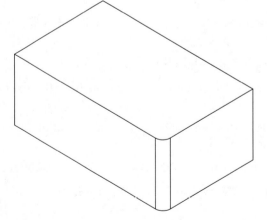

11. Save the file.

Linear Fillet

A linear fillet has a different starting and ending radius. The fillet will blend from the starting to the ending radius in a straight line. The value of zero is valid for Linear fillet. Issue the Desktop Fillet command (AMFILLET), select Linear, select Apply, and select the edge you want to fillet. Only one edge can be filleted at a time with Linear fillet. At both ends of the selected edge there will appear a "R=0". Pick one of these and type in a value and press [↵]; do the same for the other side. After you enter the second value, the fillet will be created. A linear fillet cannot be created around an edge that is already filleted or round. Figure 4.37 shows a linear fillet with the back edge set to a radius of zero and the front edge set to .75. The front edge marked with an X could not be filleted with a linear fillet. A constant or cubic fillet could be placed on this edge instead.

Figure 4.37

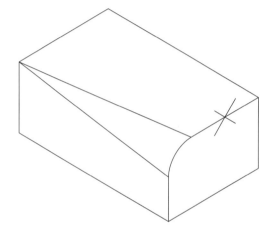

Tutorial 4.14—Creating a Linear Fillet

1. Open the file `\Md2book\Drawings\Chapter4\Ex4-14.dwg`.
2. Issue the Desktop Fillet command (AMFILLET).
3. Select Linear, return to dialog box and then select Apply.
4. Select the front left horizontal edge and the select the back R=0 and give it a value of .75 and give the front R=0 a value of 0, as shown if Figure 4.38.
5. Press [↵] to accept these values.
6. **Uncheck** Return to Dialog Box and then select Apply to create another linear fillet.

Figure 4.38

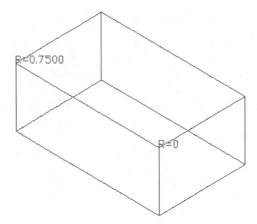

7. Select the back right horizontal edge and pick the back R=0 and give it a value of .125 and the front R=0 to .5, as shown in Figure 4.39.

Figure 4.39

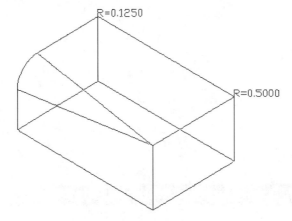

8. Press [↵] to accept these values. Your drawing should look like Figure 4.40.
9. Save the file.

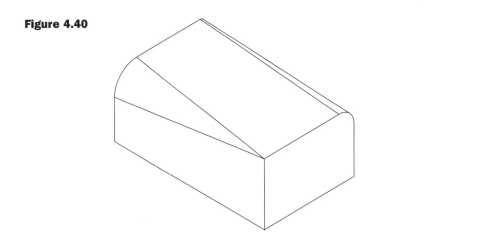

Figure 4.40

Cubic Fillet

A cubic fillet has a different starting and ending radius. The fillet will blend from the starting to the ending radius as a smooth transition, like a cubic spline. The value of zero is NOT valid for cubic fillet; the smallest fillet size is .0001. Issue the Desktop Fillet command (AMFILLET), select Cubic, select Apply, and select the edge you want to fillet. Only one edge can be filleted at a time. A "R=*" will appear at each vertex of the edge and at the command line you will see options:

```
Add vertex/Clear/Delete vertex/<Select radius>:
```

Add Vertex: Allows you to add a vertex and specify a radius. Select on the edge near where you want to add a different radius and then specify by a percentage value between the bounding vertices where you want this vertex added.

Clear: Removes the value of a selected fillet (and then you can select the "R=*" and type in a new value) or it will use the radius from the previous vertex. Delete vertex: Removes the selected vertex from the part.

Select radius: To select a radius, pick one of the R=* and type in a value and press ⏎. Do the same for any vertex that was added. After entering all the values, press ⏎ and the fillet will be created. If an R=* is not given a value, the value from the proceeding vertex will be used. Figure 4.41 shows a cubic fillet with the back edge set to a radius of .001 and the front edge set to .5. The front edge marked with an X could be filleted with a constant or cubic fillet.

Figure 4.41

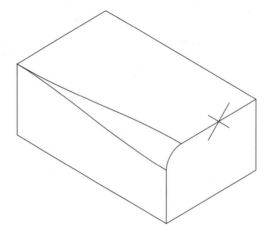

 Tutorial 4.15—Creating a Cubic Fillet

1. Open the file \Md2book\Drawings\Chapter4\Ex4-15.dwg.
2. Issue the Desktop Fillet command (AMFILLET).
3. Check Cubic and then pick Apply.
4. Select the front left horizontal edge.
5. At the command line press **A** to add a vertex and press [←Enter].
6. Select near the midpoint of the edge.
7. Type "50" so that the vertex is in the middle of the edge, and an R=* will appear at the midpoint of the edge.
8. Give a value of .125 for each end and .5 for the middle R=* as shown in Figure 4.42.
9. Press [←Enter] to accept these values. Your drawing should look like Figure 4.43.
10. Save the file.

Figure 4.42

Figure 4.43

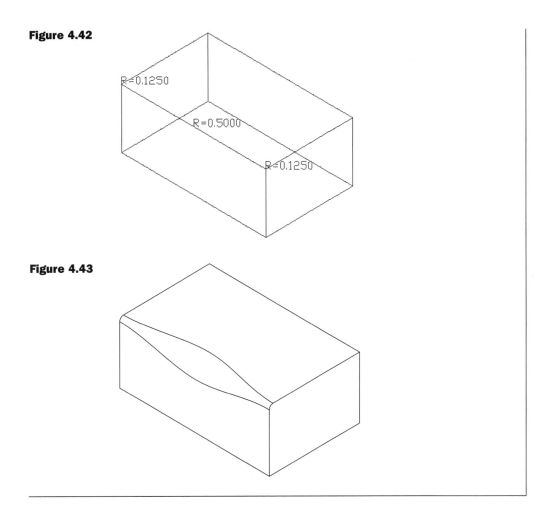

Fixed Width Fillet

A fixed width fillet is used when a fillet is created between an angled face and an extrusion that is NOT perpendicular to that face. The fixed width is the chord length between the ends of the fillet. If you drew an arc, then found the distance between the two end points of the arc, this would be the chord length. The chord length will be the same as it goes around the edge. Issue the Desktop Fillet command (AMFILLET), select Fixed Width, type in a value for Chord length, select Apply, select the edge you want to fillet, and the fillet will be created. Only one edge can be filleted at a time.

Tutorial 4.16—Fixed Width Filleting

1. Open the file \Md2book\Drawings\Chapter4\Ex4-16.dwg.
2. Issue the Desktop Fillet command (AMFILLET).
3. Select Constant, type in a radius of .125 and then select Apply.
4. Select the edge where the leftmost cylinder joins the angled plane.
5. Press [←Enter] to accept this selection. Your drawing should look like Figure 4.44. The chord length is not the same around the edge.

Figure 4.44

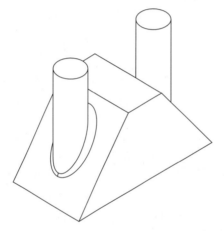

6. At the command line, change to a different isometric view by pressing **8** and [←Enter].
7. Issue the Desktop Fillet command (AMFILLET).
8. Select Fixed Width, type in a Chord Length of .125 and then select Apply.
9. Select the edge where the right cylinder joins the angled plane. Your drawing should look like Figure 4.45; the chord length is the same around the edge.
10. Save the file.

Tips: If you get an error when creating or editing a fillet, try to create it with a smaller radius.

If you still get an error when creating a fillet, you can try to create the fillets in a different sequence or to create multiple fillets in the same command.

Figure 4.45

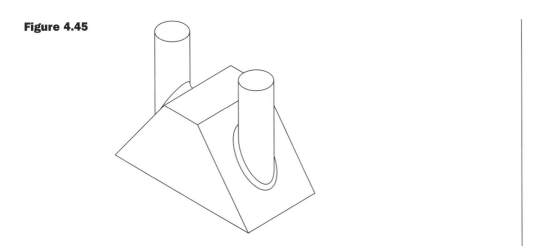

Chamfers

To create a 3D chamfer, you will select the common edge and the chamfer will be created between the two faces sharing this edge. Issue the Desktop Chamfer command (AMCHAMFER) from the Part Modeling toolbar and a dialog box will appear. Here you will choose from three operations: Equal Distance, Two Distances and Distance x Angle. After selecting an operation, type in the required information and then select OK. You will be returned to the drawing where you will select the edge(s) that you want chamfered. A description of the three operations follows.

Figure 4.46

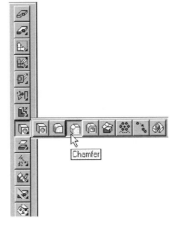

Equal Distance Chamfer

The equal distance option will create a 45° chamfer on the selected edge. The size of the chamfer is determined by the distance that you specify in the dialog box. The value is then offset in from the two common faces. From the dialog box, select Equal Distance, type in a value for the Distance1, and then select OK. Select the common edge and a 45° chamfer will be created. Multiple edges can be selected in a single command; these chamfers are linked together—if one changes or gets deleted, they all get changed or deleted.

Tutorial 4.17—Creating a Chamfer: Equal Distance

1. Open the file \Md2book\Drawings\Chapter4\Ex4-17.dwg.
2. Issue the Desktop Chamfer command (AMCHAMFER).
3. Select Equal Distance.
4. Give a value of .5 for Distance1.
5. Select OK.
6. Select the top left horizontal edge and then press [↵Enter]. When complete, your drawing should look like Figure 4.47.

Figure 4.47

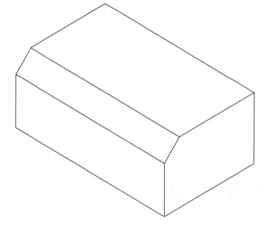

7. Press [↵Enter] to repeat the Desktop Chamfer command.
8. Select Equal Distance.
9. Type in a value of .25 for Distance1.
10. Select OK.

Cut, Join and Intersect Operations, Browser Actions, Fillets, Chamfers, Holes and Arrays

11. Select all the remaining outside edges from the existing chamfer, including the 45° edges, and then press [⏎]. Figure 4.48 shows the completed part rotated to better show the chamfers.

Figure 4.48

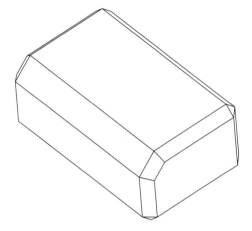

12. Save the file.

Two Distances Chamfer

The Two Distances option will create a chamfer offset in from two faces, each the amount that you specify. In the dialog box select Two Distances and give a value for both Distance1 and Distance2 (remember which distance is the first distance) and then select OK. Select the edge that you want to chamfer and a face will be highlighted. You have the option to accept this as the first face or press **N** and [⏎] to highlight the other face. The highlighted face represents the first face that Distance1 will be applied to. The distance will be offset from this face. When the correct face is selected, press [⏎] to create this chamfer. Only one edge can be chamfered at a time with this operation. After you create the chamfer, if it is the reverse of what you were expecting, edit the chamfer and the same dialog box will appear that was used to create it. Reverse the number in the dialog box and update the part.

Mechanical Desktop 2.0: Applying Designer and Assembly Modules

 Tutorial 4.18—Creating a Chamfer: Two Distances

1. Open the file \Md2book\Drawings\Chapter4\Ex4-18.dwg.
2. Issue the Desktop Chamfer command (AMCHAMFER).
3. Select Two Distances.
4. Give a value of .125 for Distance1 and 1 for Distance2.
5. Select OK.
6. Select the top left horizontal edge as shown with an X in Figure 4.49, and press ⏎ to accept this edge.

Figure 4.49

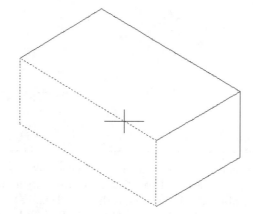

7. Press ⏎ to accept the left vertical face as highlighted in Figure 4.49.
8. When complete, your drawing should look like Figure 4.50.

Figure 4.50

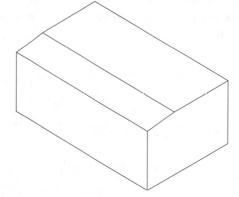

9. Create three identical chamfers for the three edges, as marked with an X in figure 4.51. The first face should be the vertical face, as highlighted in Figure 4.49 or 4.51. When complete, your part should look like Figure 4.52.

Figure 4.51

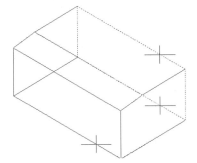

Figure 4.52

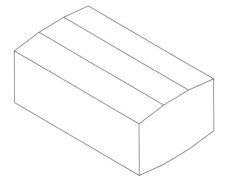

10. Create a .25 equal distance chamfer around all eight front edges (they can all be selected in a single command). When complete, your part should look like Figure 4.53.

Figure 4.53

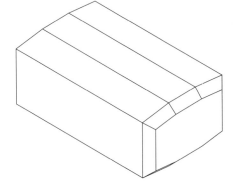

11. Save the file.

Distance x Angle Chamfer

The Distance x Angle option will create a chamfer offset down from a face and angled in from the face the number of degrees specified. In the dialog box, select Distance x Angle and give a value for both the Distance1 and the Angle and then select OK. Select the edge that you want to chamfer and a face will be highlighted. You have the option to accept this as the first face or press **N** and [⏎] to highlight the other face. The highlighted face represents the first face that Distance1 will be applied to. The distance will be offset down this face. When the correct face is selected, press [⏎] to create this chamfer. Only one edge can be chamfered at a time with this operation.

Tutorial 4.19—Creating a Chamfer: Distance x Angle

1. Open the file \Md2book\Drawings\Chapter4\Ex4-19.dwg.
2. Issue the Desktop Chamfer command (AMCHAMFER).
3. Select Distance x Angle.
4. Give a value of 1 for Distance1 and 60 for the angle.
5. Select OK.
6. Select the top right horizontal edge as shown with an X in Figure 4.54, and press [⏎] to accept this edge.
7. Press [⏎] to accept the right vertical face as shown in Figure 4.54. When complete, your part should look like Figure 4.55.

Figure 4.54

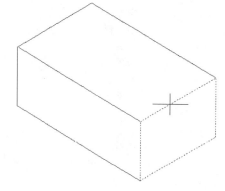

Figure 4.55

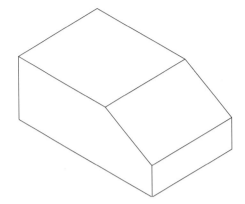

8. Press [⏎] to create the same type and size chamfer on the front bottom edge. The same front face will be used for the offset, as highlighted in Figure 4.54. When complete, your part should look like Figure 4.56.

Figure 4.56

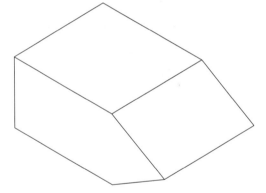

9. Save the file.

Holes

There are thee basic types of holes that Mechanical Desktop can create: drilled, counter bore and counter sink. Mechanical Desktop can also simulate a tapped hole, but it will only show up in the drawing views, appearing as a drilled hole in the part. When editing a hole, use either the Edit Feature command (AMEDITFEAT) or the browser, and the same dialog box will appear that created it. Make any necessary changes and select Apply. If dimensions were used to place the hole, they will reappear on the screen. To change a dimension, select it, type in a new value, and press [⏎]. Select another dimension to change or press [⏎] to exit the command. Update the part if it did not automatically update.

Mechanical Desktop 2.0: Applying Designer and Assembly Modules

To create a hole, issue the Hole command (AMHOLE) from the Part Modeling toolbar and a dialog box will appear. The hole dialog box is broken into six areas: Operation, Termination, Placement, Drill Size, C'Bore/Sunk Size and Tapped. A description follows of the different areas.

Figure 4.57

Figure 4.58

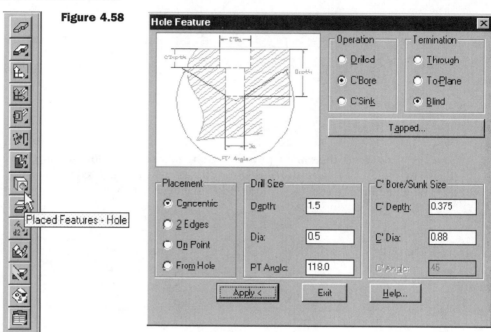

Operation

Option	Function
Drilled	Drilled hole.
C'Bore	Counter bore hole.
C'Sink	Counter sink hole.

Termination

Option	Function
Through	Will go all the way through the part.
To-Face	Will stop at a selected face.
Blind	Will stop at a specified depth.

Placement

Option	Function
Concentric:	With this option you will select a plane in which the hole should start and then select an arc or circular edge to which the hole will be concentric.
2 Edges	Select two edges that define a plane, then select a point near where the hole should be. You will then be prompted for the exact offset distance from each edge.
On Point	The hole will be placed on a selected work point.
From Hole	You will be prompted to select a plane to place a hole and then you will select a hole to offset for both the X and Y direction.

Drill Size

Option	Function
Depth	The depth for a blind hole.
Dia	The diameter for the drilled hole.
PT Angle	The point angle for a blind hole.

C'Bore/Sunk Size

Option	Function
C'Depth	Depth for the counter bore.
C'Dia	Diameter for the counter bore or sink.
C'Angle	The angle for the counter sink.

Tapped

Tapped holes are represented in the part as drilled holes. When the drawing views are created they will be represented by the drafting standard that you select through the drawing preferences. To create a tapped hole, first select Drilled hole, type in a diameter for the hole, select a termination and then select Tapped in the dialog box. A second dialog box will appear; select Tapped in the upper left corner of the dialog box and type in a value for the tapped diameter. Select Full Depth if the tap should be the same depth as the drilled hole; otherwise type in a value for the depth of the tap. Select OK and you will be returned to the Hole Feature dialog box. Select Apply to create the hole.

Figure 4.59

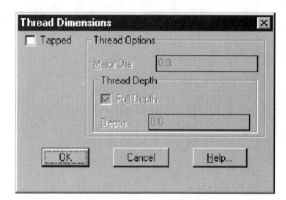

Tutorial 4.20—Creating a Hole: Drilled, Through, Blind, Concentric and Two Edges

1. Open the file \Md2book\Drawings\Chapter4\Ex4-20.dwg.

2. Issue the Hole command (AMHOLE).

3. Select Drilled, Blind, Concentric, type ".5" for the Diameter, ".75" for the Depth and then select Apply.

4. For the prompt:

   worldXy/worldYz/worldZx/Ucs/<Select work plane or planar face>

   select a point in the plane as highlighted in Figure 4.60.

5. For the prompt:

   Select concentric edge:

   select either the top or bottom circle defining the larger cylinder.

Figure 4.60

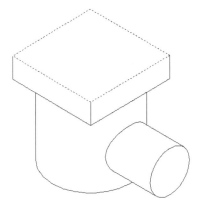

6. Press [⮐] to return to the dialog box and move the dialog box to the side of the screen. Your part should look like Figure 4.61.

Figure 4.61

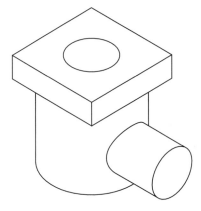

7. In the dialog box, select Drilled, Through, Concentric, type ".25" for the Diameter and then select Apply.

8. For the prompt:

   worldXy/worldYz/worldZx/Ucs/<Select work plane or planar face>

   select on the circumference of the outside boss, as shown with an X in Figure 4.62.

Figure 4.62

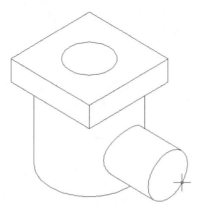

9. For the prompt:

   ```
   Select concentric edge:
   ```

   select the same circle as in step 8.

10. Press [⏎] to return to the dialog box and move the dialog box over. Your part should look like Figure 4.63.

Figure 4.63

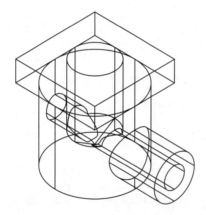

11. In the dialog box select Drilled, Through, Two Edges, type ".125" for the Drill Diameter and then select Apply.

12. Select the two highlighted edges and then select a point where the two part lines intersect, as shown in Figure 4.64.

Figure 4.64

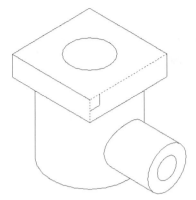

13. At the command line, type ".125" for both the distances from the first and the second edge.

14. While still in the command, select the two opposite edges and give them ".125" for each distance. Your part should look like Figure 4.65.

Figure 4.65

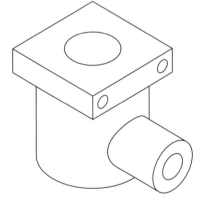

15. Save the file.

Mechanical Desktop 2.0: Applying Designer and Assembly Modules

Tutorial 4.21—Creating a Hole: Counter Bore, Counter Sink, Through, Concentric and On Point

In this tutorial, the work points have already been created and dimensioned.

1. Open the file \Md2book\Drawings\Chapter4\Ex4-21.dwg.

2. Issue the Hole command (AMHOLE).

3. Select Counter Bore, Through, On Point, type ".14 " for the Drill Diameter, ".14" for the counter bore depth and ".25" for the counter bore diameter. Then select Apply.

4. Select both work points and then press [↵Enter] to return to the dialog box and move the dialog box over. Your part should look like Figure 4.66.

Figure 4.66

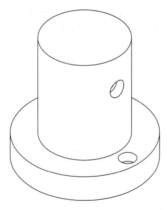

5. Select Counter Sink, Through, Concentric, type ".5" for the Drill Diameter, "1" for the Counter Sink Diameter, "45" for the Counter Sink Angle and then select Apply.

6. Select on the circumference of the top cylinder twice, press [↵Enter] to return to the dialog box and select Exit to finish the command. When complete, your part should look like Figure 4.67.

7. Save the file.

Figure 4.67

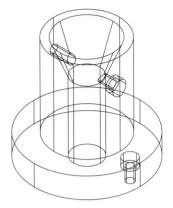

Tutorial 4.22—Creating a Hole: Drill, Two Edges, From Point and Editing

1. Open the file \Md2book\Drawings\Chapter4\Ex4-22.dwg.

2. Issue the Hole command (AMHOLE).

3. Select Drilled, Through, Two Edges, type ".25" for the Drill Diameter and then select Apply.

4. Select the two highlighted edges and then select a point near the intersection of the two lines, as shown in Figure 4.68.

Figure 4.68

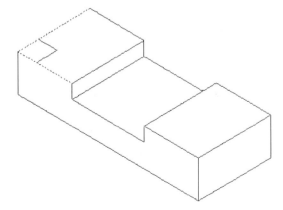

5. At the command line, type in a value of .5 for distance from both the first and the second edge.

6. Press [Enter] to return to the dialog box and select From Hole, change the Drill Diameter to .5 and then select Apply.

7. At the prompt:

    ```
    worldXy/worldYz/worldZx/Ucs/<Select work plane or planar face>:
    ```

    select the highlighted plane and rotate the UCS as shown in Figure 4.69.

Figure 4.69

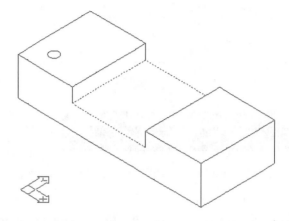

8. At the prompt:

    ```
    Select x direction reference hole:
    ```

    select the hole that was just created.

9. At the prompt:

    ```
    Select y direction reference hole/<Previous>:
    ```

    either select the same hole or press [Enter] to use the previously selected hole.

10. Select a point near the middle of the part, as shown in Figure 4.70. As you move the mouse, you are shown graphically the approximate position of the hole.

11. At the prompt:

    ```
    Enter x distance <1.881393>:
    ```

    type "2" and press [Enter].

12. At the prompt:

    ```
    Enter y distance <0.560419>:
    ```

    type ".5" and press [Enter]. Your part should look like Figure 4.71.

Figure 4.70

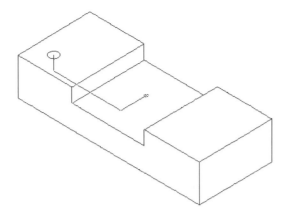

Figure 4.71

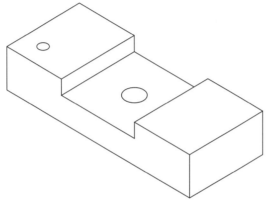

13. Press to return to the dialog box. Change the Drill Diameter to .25 and then select Apply.

14. At the prompt:

    worldXy/worldYz/worldZx/Ucs/<Select work plane or planar face>:

    select the highlighted plane and rotate the UCS as shown in Figure 4.72.

15. At the prompt:

    Select x direction reference hole:

    select the hole in the lower left corner.

16. At the prompt:

    Select y direction reference hole/<Previous>:

    select the hole in the middle of the part.

Figure 4.72

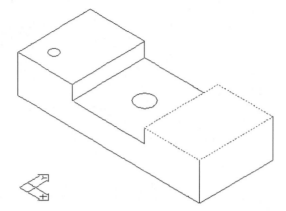

17. Select a point near the top right corner of the part, as shown in Figure 4.73. As you move the mouse, you are shown graphically the approximate location of the hole.

Figure 4.73

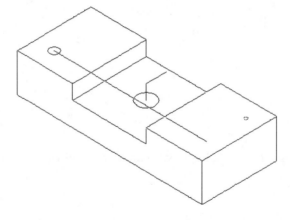

18. At the prompt:

    Enter x distance <3.872356>:

    type "4" and press ⏎.

19. At the prompt:

    Enter y distance <0.480586>:

    type ".5" and press ⏎ to return to the dialog box. Select Exit to finish the command. Your part should look like Figure 4.74.

20. Edit the last hole created with the Edit Feature command (AMEDITFEAT) or use the browser to edit hole3. Change the drill diameter to .5 and Select Apply, change the dimensions from "4" to "3.5" and the ".5" to "0" and press ⏎.

Figure 4.74

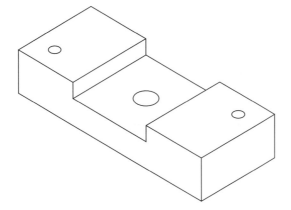

21. Update the part if necessary. Your part should look like Figure 4.75.

Figure 4.75

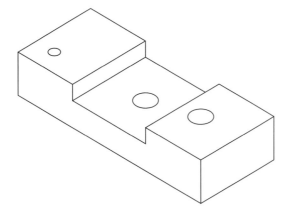

22. Save the file.

Arrays

In Mechanical Desktop, there are two type of arrays, rectangular and polar. Before creating a polar array, you must have a work point or work axis around which the array will be rotated. Both the rectangular and polar arrays are accessed from the same command, Feature Array (AMARRAY), found on the Part Modeling toolbar as shown in Figure 4.76. After issuing the command, you are prompted to select a feature to array. After you select a feature, a dialog box will appear, and if you move the dialog box away from the part, you will see an image that shows the direction of rows, columns and the positive angle for the selected feature (see Figure 4.77). In the dialog box, select either the rectangular or polar tab.

Rectangular Arrays

Under the rectangular tab, you will input the number of rows, columns and the spacing between each. The spacing can be a negative value; this will force the features to be arrayed in the negative direction.

Polar Arrays

Under the Polar tab, you have three array types, Full Circle, Spacing and Incremental Angle.

Full Circle: Type in a value for the Number of Instances and that number will be evenly spaced around 360°.

Included Angle: Type in a value for the Number of Instances and the Angle. The number of instances will be evenly spaced within the angle specified.

Incremental Angle: Type in a value for the Number of Instances and the Angle. Each instance will be separated the number of degrees specified by the value of the angle.

When checked, the Rotate as Copied option allows the features to be rotated as they are copied; otherwise they will maintain the orientation of the selected feature. After filling in the information, select OK and then select a work point or work axis as the point of rotation, and the command will be complete.

Editing Arrays

For both rectangular and polar arrays, after the feature is arrayed, the arrayed features or instances have a child relationship to the feature that was arrayed—the parent. If the size of the parent feature changes, all the child features will also change. If a hole is arrayed, and the parent hole type changes, the child holes will also change. You can edit the arrayed set by using the Edit Feature command (AMEDITFEAT) or through the browser. After issuing the command, select any of the arrayed holes. Then you can change the number of rows, columns, or arrayed features, the spacing of the rows and columns and the angle for polar arrays. After the arrayed set is selected, the dimensions or an angle will appear on the selected feature, as well as "C#=" . These "C#=" represent the number of rows, columns or number of polar arrayed features. These "C#=" will be incremented by one as each new array is created. To edit or delete an arrayed feature by itself, issue the Edit Feature command (AMEDITFEAT) or select the array from the browser and press I to edit the selected feature independently of the set. Then select the feature. Once a feature is independent of the array, it has no relationship to the arrayed set and it can be edited or deleted independently. If the number of rows or columns of the arrayed set changes, the location of the independent feature will be maintained. This feature will not be constrained to the part. To constrain the feature, use the Edit Feature command with the Sketch option, as described in Chapter 2.

Figure 4.76

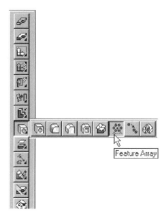

Figure 4.77

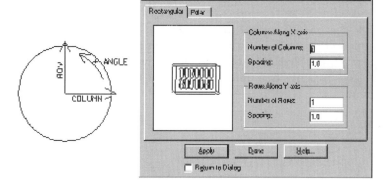

Notes: A base part or feature cannot be arrayed.

For polar arrays, a work point or work axis must be created before you issue the Feature Array command (AMARRAY); either one can be used as the axis of rotation.

Tutorial 4.23—Creating and Editing a Rectangular Array

1. Open the file \Md2book\Drawings\Chapter4\Ex4-23.dwg.
2. Issue the Feature Array command (AMARRAY).
3. Select the hole in the plate.
4. Type "5" for the Number of Columns, "1" for the column Spacing, "2" for the Number of Rows and "2" for the row Spacing and then select Apply. When complete, your part should look like Figure 4.78.

Figure 4.78

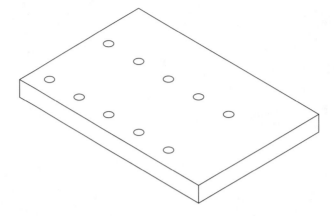

5. Change to the plan view, the **9** key.

6. Issue the Edit Feature command (AMEDITFEAT) command or from the browser, double-click on RectArray1. If you used the icon to issue the command, select any one of the arrayed features. Then the dimensions for the spacing will appear on the selected feature along with two C#=#, as shown in Figure 4.79. The C0 represents the number of rows and C1 represents the number of columns.

Figure 4.79

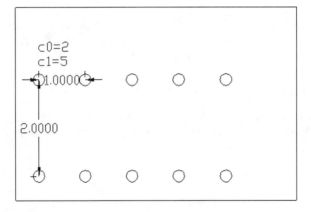

7. Change the number of rows by selecting C0=2 and typing in "4" and then press [⏎]. Change the number of columns by selecting C1=5 and typing in "6" and thenpress [⏎]. Change the row spacing from "2" to "1" and press [⏎]. Update the part if necessary. When complete, your part should look like Figure 4.80.

8. Press [⏎] to repeat the Edit Feature command.

9. Press I and [⏎] to edit the Independent array instance.

Cut, Join and Intersect Operations, Browser Actions, Fillets, Chamfers, Holes and Arrays

Figure 4.80

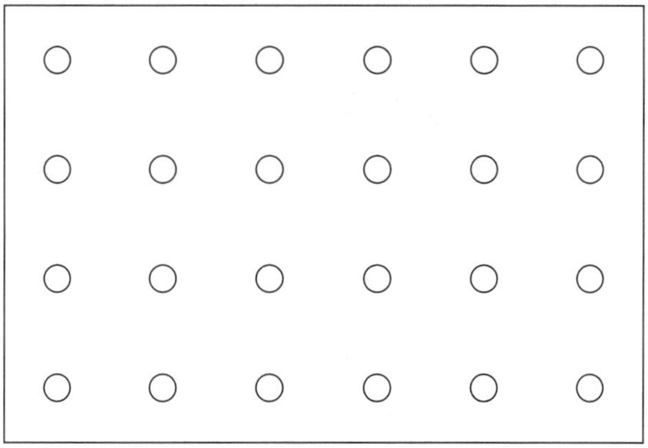

10. Select the hole in the upper right corner and you will be returned to the command prompt.
11. Press [⏎] to repeat the Edit Feature command.
12. Select the hole in the upper right corner and change its diameter to .375; update the part.
13. Delete the hole in the upper right corner. Your part should look like Figure 4.81.

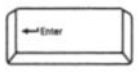

Figure 4.81

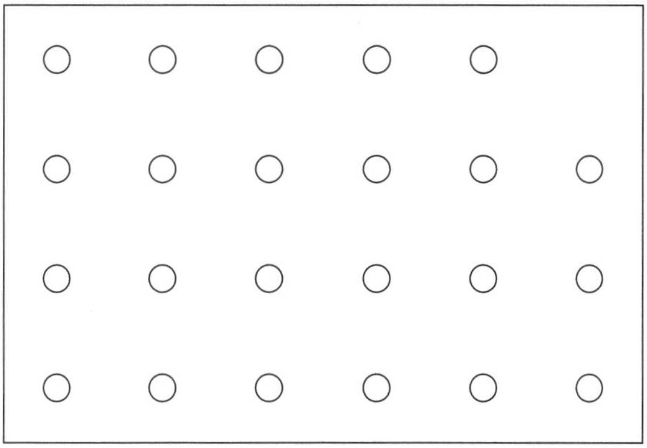

14. Save the file.

Tutorial 4.24—Creating a Polar Array: Full Circle

1. Open the file \Md2book\Drawings\Chapter4\Ex4-24.dwg.
2. Issue the Feature Array command (AMARRAY) and select the counter bore hole on the bottom of the part.
3. Select the Polar tab and type "3" for the Number of instances, select Full Circle, check Rotate as Copied and then pick OK.
4. Select the work axis and your part should look like Figure 4.82.

Figure 4.82

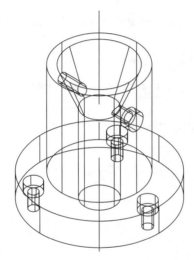

5. Save the file.

Tutorial 4.25—Creating and Editing a Polar Array: Included Angle

1. Open the file \Md2book\Drawings\Chapter4\Ex4-25.dwg.
2. Issue the Feature Array command (AMARRAY) and select the triangular extrusion.
3. Select the Polar tab and type "3" for the Number of instances, select Included Angle and type "-180" for the Angle, check Rotate as Copied and then select OK.

4. Select the work point and your part should look like Figure 4.83.

Figure 4.83

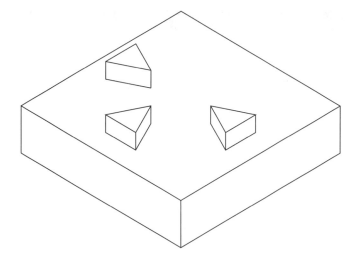

5. Issue the Edit Feature command (ANEDITFEAT) or through the browser, edit the polar array PolarArray1. Select one of the arrayed features. A dimension for the angle will appear, along with a C0=3. Change the C0=3 to 4 and the -180 to -220. Update the part if necessary. When complete, your part should look like Figure 4.84.

Figure 4.84

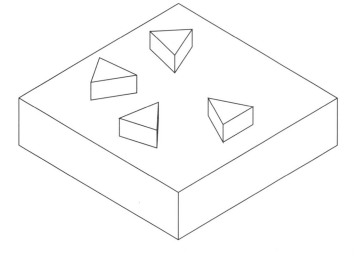

6. Save the file.

Tutorial 4.26—Creating a Polar Array: Included Angle

1. Open the file \Md2book\Drawings\Chapter4\Ex4-26.dwg.

2. Issue the Feature Array command (AMARRAY) and select the triangular extrusion.

3. Select the Polar tab and type "3" for the Number of instances, check Included Angle, type "60" for the Angle, check Rotate as Copied and then select OK. Select the work point and your part should look like Figure 4.85.

Figure 4.85

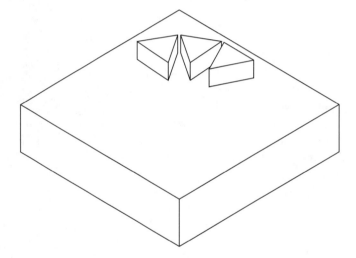

4. Save the file.

Exercises

Open and then complete the following three exercises, as shown in the finished drawing. When each is complete, save the file.

Exercise 4.1—Bracket

Open the file \Md2book\Drawings\Chapter4\bracket.dwg.

Figure 4.86

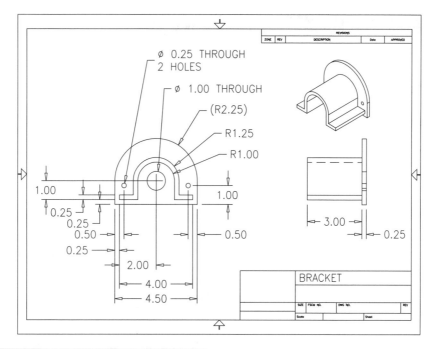

Exercise 4.2—Guide

Open the file \Md2book\Drawings\Chapter4\Guide.dwg.

Figure 4.87

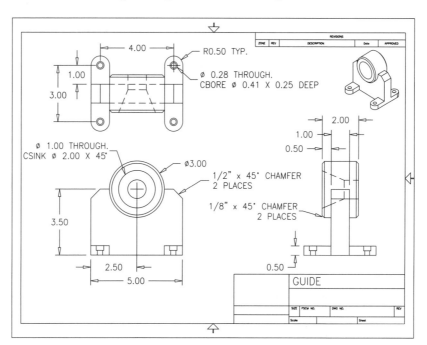

Mechanical Desktop 2.0: Applying Designer and Assembly Modules

Exercise 4.3—Foot

Open the file `Md2book\Drawings\Chapter4\Foot.dwg`.

Figure 4.88

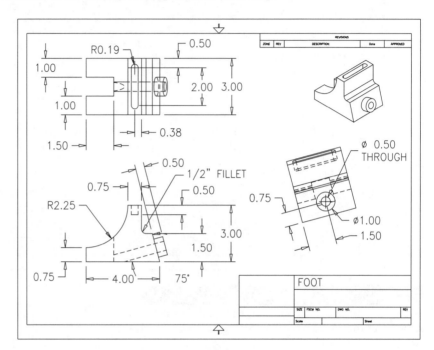

Cut, Join and Intersect Operations, Browser Actions, Fillets, Chamfers, Holes and Arrays

Review Questions

1. What are the three different operation types used for sketched features?
2. An operation can be performed without having a base part. T or F?
3. When a feature is extruded Through a part and the depth of the part increases, where the feature was extruded, the feature will no longer go through the part. T or F?
4. What is the difference between a sketch plane and a work plane?
5. You are always required to create a work plane before drawing a sketch. T or F?
6. What is the difference between a linear and a cubic fillet?
7. You can create a cubic fillet on multiple edges by picking the edges in a single command. T or F?
8. Name two ways to edit an existing hole.
9. After you create a rectangular array, how can one of the arrayed features be edited?
10. A polar array can be arrayed about the center of an arc or circle with a selection on the circumference of the arc or circle. T or F?
11. A feature cannot be copied from one part to another. T or F?

chapter 5

Advanced Dimensioning, Constraining and Sketching Techniques

In this chapter you will learn how to get more out of Mechanical Desktop by using advanced techniques. You will learn how to set up relationships between geometries, use existing edges to close a sketch, use construction geometry to better control the sketch and also how to convert existing 2D AutoCAD drawings to 3D parametric parts.

After completing this chapter, you will be able to:

- Create relationships between dimensions.
- Create both local and global variables.
- Create a table driven by an Excel spreadsheet.
- Create a profile and use existing geometry to close the profile.
- Use construction geometry to help constrain profiles.
- Convert existing 2-D drawings to parametric sketches.

Dimension Display and Equations

When creating parts, you may want to set up a relationship between dimensions. For example, the length of a part may need to be twice that of its width. In Mechanical Desktop, each dimension is given a label. Each label starts with the letter "d" and is given a number, for example "d0" or "d27". The first dimension created is given the label "d0" and each dimension that follows sequences up one number at a time. If a dimension is erased, the next dimension does **not** go back and reuse the erased value. Instead, it keeps sequencing from the last value on the last dimension created. To change the display of the dimensions so that you can see what these "d" values are, issue the Dimension Display command (AMDIMDSP), as shown in Figure 5.1.

There are three different dimensional display modes: parameters, equation and numeric. After you select a dimension display mode, all the dimensions on the screen will change to that mode.

Parameters mode: Displays as the dimension number or "d#".

Equation mode: All the dimensions on the screen will change to "d# = #", showing each actual value, for example, d7=4.50.

Numeric mode: The default. Displays the dimensions as actual numbers.

As dimensions are created, they will reflect the current dimension display mode, which has no effect on the dimension's value. To create a dimension using another dimension's "d#", type in the "d#" when prompted to verify a dimension's value. You can also use mathematical operators like "(d9/4)*2" or "65/9". Figure 5.2 shows all the mathematical operators that Mechanical Desktop supports. Use the Change Dimension command (AMMODDIM) from the Part Modeling toolbar to add or remove a relationship between dimensions.

Note: After a dimension with a parameter is created, it can be changed with the Change Dimension command.

Tip: When creating relationships between dimensions, use the equation display mode to see the "d#" and the actual value of the dimension.

Figure 5.1

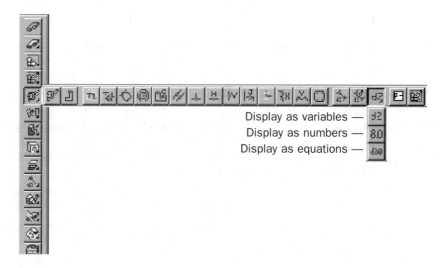

Display as variables
Display as numbers
Display as equations

Figure 5.2

Operator	Description
^	Exponent
+	Add
-	Subtract
*	Multiply
/	Divide
%	Modulus (remainder)
sqrt	Square root
log	Logarithm
ln	Natural Logarithm
floor	Rounds down to nearest whole number
ceil	Rounds up to nearest whole number
sin	Sine
cos	Cosine
tan	Tangent
asin	Arcsin ($\sin^{-1}$)
acos	Arcos ($\cos^{-1}$)
atan	Arctang ($\tan^{-1}$)
sinh	Hyperbolic Sine
pi	Pi

Mechanical Desktop 2.0: Applying Designer and Assembly Modules

Tutorial 5.1—Using Dimension Display and Equations

1. Open the file \Md2book\Drawings\Chapter5\Ex5-1.dwg.
2. Profile the geometry.
3. Change the Dimension Display (AMDIMDSP) mode to Parameters (tooltip refers to the modes as variables).
4. Create the right vertical dimension with the value of "1". When complete, your drawing should look like Figure 5.3.

Figure 5.3

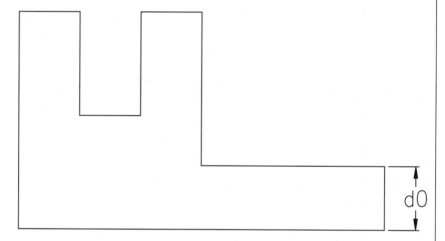

5. Create the three horizontal dimensions, dimension them left to right and type in "d0" for each value. When complete, your drawing should look like Figure 5.4. If they were dimensioned in a different order, the "d#" would be different; the order in which the dimensions are placed will have no effect on the actual part.
6. Change the Dimension Display mode to Equations.
7. Add two vertical dimensions on the left side of the part. Place the inside vertical line as "d0*2" and the outside vertical line as "d0*3". When complete, your drawing should look like Figure 5.5. The three horizontal dimensions have been repositioned for clarity.
8. Dimension the bottom horizontal line with the equation "(d5*2)-d0". When complete, your drawing should look like Figure 5.6.
9. Change the Dimension Display mode to Numeric. When complete, your drawing should look like Figure 5.7.

Figure 5.4

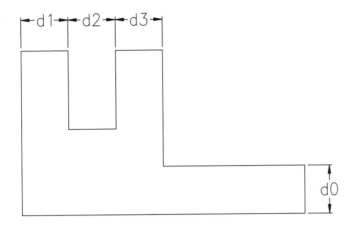

Figure 5.5

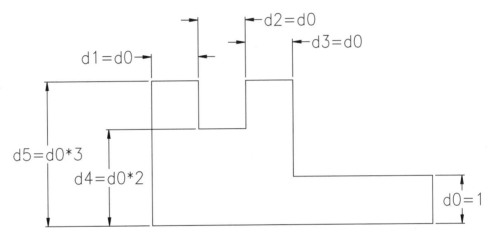

Figure 5.6

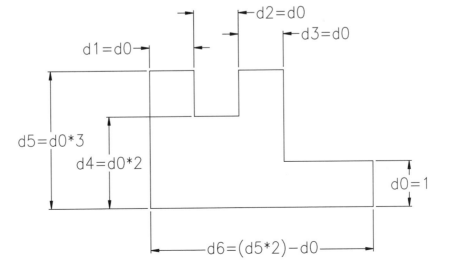

Figure 5.7

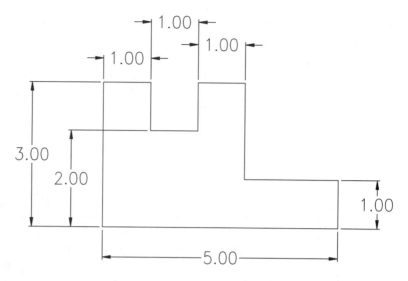

10. Change to an isometric view, the **8** key.

11. Extrude the profile; in the extrusion distance, type in "d0". When complete, your part should resemble Figure 5.8.

Figure 5.8

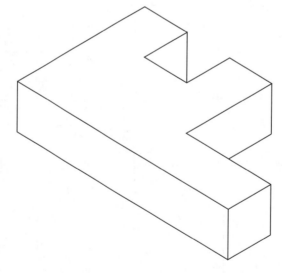

12. Edit the part and change the "5" bottom horizontal dimension "(d5*2)-d0" to "4", and change the "d0" (the vertical dimension created first) to "2". Update your part and your drawing should look like Figure 5.9.

13. Save the file.

Figure 5.9

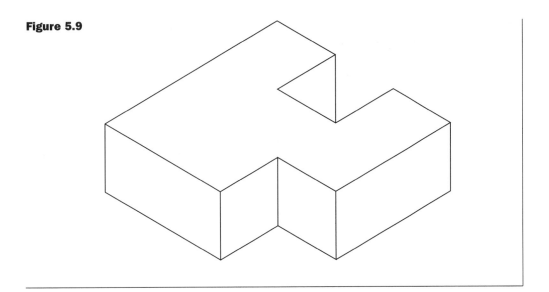

Design Variables

A design variable is a user-defined name that is assigned numeric value either explicitly or through equations. A design variable can be used anywhere a dimension is required. A design variable can be defined by any name except "c# or d#" where # is any number (those letter-number combinations are used for arrays and dimensions). There are two types of design variables: active part and global. Active part design variables can be used in the active part only. Global design variables can be used in multiple parts in the same file as well as in assembly constraints. If an active and a global design variable exist with the same name in the same file, the active variable takes precedence over the global variable.

Active Part Variables

To create an active part design variable, you must first have a sketch that has been profiled. After a sketch has been profiled, it becomes the active part. Then an active part design variable can be created. Issue the Design Variables command (AMVARS) from the Part Modeling toolbar, as shown in Figure 5.10, or right-click on the part name in the browser and choose Design Variables from the pop-up menu. After the command is issued, a dialog box will appear, as shown in Figure 5.11. There are two tabs: Active Part and Global. To create an Active Part variable, select New and a second dialog box will appear. Type in information for the name, equation and comment fields and then select OK.

Name: Can contain no spaces, can be up to 72 characters in length and must start with a letter. However, it is recommended that you keep the names short.

Equation: Any mathematical operator can be used, as described in Figure 5.2. Other design variables can also be used in the equation.

Comment: Place a description of up to 250 characters with spaces. The comment field is used only for the operator's reference and does not need to be filled in.

After filling in the information, select OK and you will be returned to the Design Variables dialog box. The information that was filled in will appear and the value field will reflect the value of the equation. Along the top row of the dialog box are the letters T D U. They represent:

T = Variable is Table-driven.

D = Duplicated variable in the active part, global variable or table-driven. For duplicate variables, the active part variable takes precedence over global variables.

U = The variable is Unreferenced in the part(s).

The letter T, D, or U will appear on the line of the variable that matches. To edit a design variable, double-click on the area that you want to edit. Type in the new information, press [↵] and then select OK. If you edit the design variables through the browser, the part will update automatically. Otherwise, use the Update Part command (AMUPDATE) to update the part. To remove a single design variable that is not being referenced, use the Delete button on the right side of the dialog box. To remove all design variables that are not being referenced, use the Purge button on the right side of the dialog box.

Global Variables

Global design variables can be created in the same manner as Active part variables except that you select the Global tab in the dialog box and fill in the name, equation and comment. Global variables can be used in any part in the file and can be used for values in assembly constraints.

The Global tab has a section named **Global Variable File**. In this area you have the option to import, link, export and unlink global variables. When variables are exported, they are saved in an ASCII file with the extension "prm". This ASCII file can be edited with any word processor. When saving the file, make sure that it is in ASCII format.

Import... Brings design variables into the current drawing from a "prm" file. The variables become global variables in the current file.

Link... Creates a link between a "prm" file and the current drawing. The drawing gets the variables from the file. If the "prm" file changes, you can update the drawing with those changes either by reopening the drawing file or by reestablishing the link (this will bring in the new variables).

Advanced Dimensioning, Constraining and Sketching Techniques

Export ... Take the global variable in the current drawing and create a "prm" file.

Unlink Remove the reference between the "prm" file and the current file. The variables will become global variables in the file.

The other area of the Global tab, **Copy to Active part**, allows you to copy global variables into active part variables. The options are Selected, Referenced and All.

Selected: The highlighted variable will be copied to an active part variable.

Referenced: All global variables that are referenced in the active part will be copied to active part variables.

All: All global variables will be copied to active part variables.

Figure 5.10

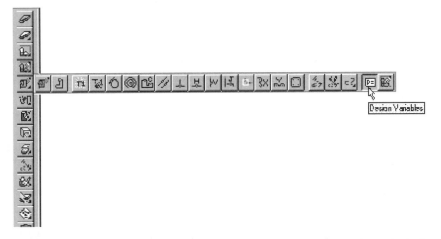

Figure 5.11

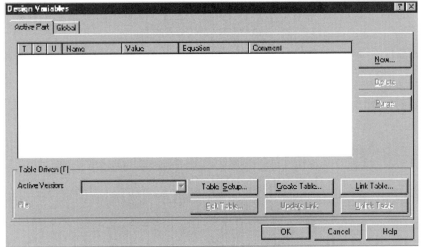

Mechanical Desktop 2.0: Applying Designer and Assembly Modules

 Note: In order for an active part variable to be created, there must be at least one sketch that has been profiled.

Active part variables take precedence over global variables!

 ## Tutorial 5.2—Using Active Part Design Variables

1. Open the file \Md2book\Drawings\Chapter5\Ex5-2.dwg.
2. Profile the sketch and use the default part name.
3. Issue the Design Variables command (AMVARS).
4. Select New and give the values Name: LENGTH, Equation: 2, Comment: length of part.
5. Select OK in the new Part Variable dialog box.
6. To create another variable, select New and give the values Name: WIDTH, Equation: LENGTH/2, Comment: width of part.
7. Select OK in the new Part Variable dialog box.
8. Select OK in the Design Variables dialog box.
9. Dimension the sketch as shown in Figure 5.12. The dimensions are shown in equation mode. Type in the corresponding variables and values.

Figure 5.12

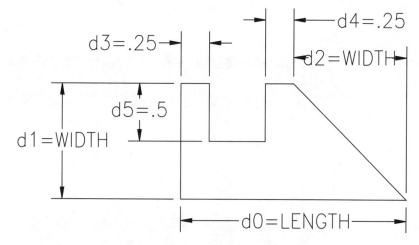

10. Change to an isometric view (the **8** key) and extrude the profile with the value "LENGTH/4" and then press [↵] to accept the default direction. Your part should resemble Figure 5.13.

Figure 5.13

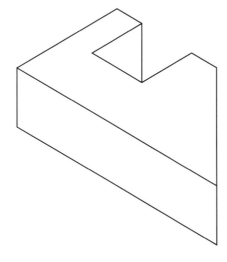

11. Edit the design variable so that LENGTH = "3" and WIDTH = "1".

12. Update the part if needed. The part should resemble Figure 5.14.

Figure 5.14

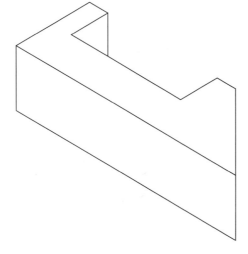

13. Save the file.

Tutorial 5.3—Using Global Design Variables

1. Open the file \Md2book\Drawings\Chapter5\Ex5-3.dwg.

2. Issue the Design Variables command (AMVARS).

3. Select the Global tab.

4. Select New and give the values Name: HoleDia, Equation: 1, Comment: Diameter of hole.

5. Select OK in the new Part Variable dialog box.

6. Select OK in the Design Variables dialog box.

7. Create a drilled through hole using the variable HoleDia for its diameter, and create two holes 1" from each edge from the lower left and upper right corner on top of the plate. When complete, the part should look like Figure 5.15.

Figure 5.15

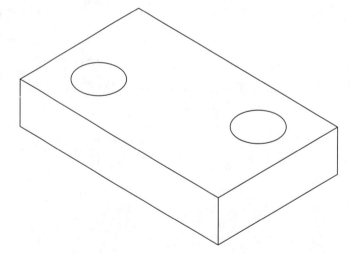

8. Issue the Design Variables command (AMVARS) and select the Global tab.

9. Select the Export option in the dialog box and save the file with the name \Md2book\Drawings\Chapter5\ **Ex5-3**. The prm extension will automatically be created when the file is created. This "prm" file will be used in the next tutorial.

10. Save the file.

Tutorial 5.4—Using Global and Linked Design Variables

1. Start a new drawing from scratch.

2. Draw a circle and profile it.

3. Issue the Design Variables command (AMVARS) and select the Global tab.

4. Select Link from the lower part of the dialog box, select the file \Md2book\Drawings\Chapter5\ Ex5-3.prm and then select OK.

5. Dimension the circle with its diameter equal to "HoleDia".

6. Extrude the circle a distance of "3" and accept the default direction.

7. Save the file as \Md2book\Drawings\Chapter5\Ex5-4.dwg.

8. Open the file "Ex5-3.prm" with your favorite word processor and change the variable "HoleDia" to ".5" and save the file. The file must be saved in ASCII text format.

9. Reopen the file \Md2book\Drawings\Chapter5\Ex5-4.dwg and update the part. Then edit the hole to verify that the new diameter is ".5".

10. Save the file.

11. Open the file \Md2book\Drawings\Chapter5\Ex5-3.dwg.

12. Issue the Design Variables command (AMVARS) and select the Global tab.

13. Select the Link option from the lower part of the dialog box, select the file \Md2book\Drawings\Chapter5\ Ex5-3.prm and then select OK.

14. If necessary, update the part. The part should resemble figure 5.16.

Figure 5.16

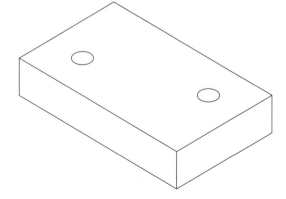

15. Save the file.

Table-driven Parts

Table-driven variables are similar to design variables, except that the information is created in an Excel Spreadsheet. The spreadsheet can have different values of the same variable if multiple rows are created under the variable name, as shown in Figure 5.17. The table-driven variables will be used as active part design variables and cannot be used as global variables or for values in assembly constraints. Table-driven variables are used in the modeling process just like design variables; they can be placed in dimensions of a sketch or existing part. If the part is using an active part design variable with the same name as a table-driven variable, the active design variable will be overridden by the table-driven-design variable. The spreadsheet is linked to the part, and changes in Excel can be updated inside Mechanical Desktop to change the part.

To create a table-driven variable, you must first have a sketch that has been profiled. From the Part Modeling toolbar, issue the Design Variables command (AMVARS) or right-click on the part name in the browser and choose Design Variable from the pop-up menu.

Figure 5.17

	A	B	C	D
1		Length	Width	Depth
2	Part A	5	4	0.25
3	Part B	7	3	0.5
4	Part C	8	7	0.75
5	Part D	12	10	1

The seven basic steps to create table-driven variables:

1. Create a sketch and profile it, then right-click on the part name in the browser and choose Design Variables from the pop-up menu or select the Design Variables command (AMVARS) from the Part Modeling toolbar.

2. Select Table Setup from the dialog box and a Table Setup dialog box will appear. Choose whether the versions should go down or across the spreadsheet and also which sheet number or sheet name this should be linked to. Select OK when complete.

3. Select Create Table and a Create Table dialog box will appear. Give the spreadsheet a name and location, select Save and Excel will load, if it is not already running. If there are active design variables, they will be written out to this Excel file with a name of Generic. Generic can be renamed to any name without spaces.

Advanced Dimensioning, Constraining and Sketching Techniques

4. In Excel, input the information as needed and then save the file. (The A1 cell is reserved and cannot be used.) Then add the data to the rows and columns of the spreadsheet.

5. Return to Mechanical Desktop and select the Update Link button. The variables will be added to the Design Variables dialog box and the versions will appear in the browser.

6. Dimension the sketch or edit the part to add the Design Variables. Finish the part as required.

7. To change between versions, you can double-click on the version in the browser or, in the Design Variables dialog box, select the version from the Active Version drop-down list.

The information in the Excel spreadsheet can be linked to a Mechanical Desktop drawing. Highlight the cells in Excel that you want linked to Mechanical Desktop, copy them to the clipboard using **Ctrl+C** or right-click on the highlighted area and then choose Copy from the pop-up menu. Switch back to Mechanical Desktop and choose Paste Special from the Edit pull-down menu and select a point to locate the chart. Now that the chart is linked to the spreadsheet, when the variable(s) change inside Excel, they will automatically be updated in the chart inside Mechanical Desktop.

To update the variables in Mechanical Desktop, either right-click on the top level of the table in the browser and choose Update from the pop-up menu or issue the Design Variables command (AMVARS) and select Update Link from the dialog box. The new values will be current. If you open a drawing and the table is red in the browser, this is a warning that there is a problem in locating the spreadsheet. Right-click on the table in the browser and choose Resolve Conflict from the pop-up menu and select the new location of the table.

Note: Table-driven parts require Excel.

Tutorial 5.5—Using Table-driven Parts as Design Variables

In this tutorial, the part is complete with active part design variables. You will create a table and then link the Excel spreadsheet to the part.

1. Open the file \Md2book\Drawings\Chapter5\Ex5-5.dwg.

2. Issue the Design Variables command (AMVARS) or right-click on Part1_1 and choose Design Variables from the pop-up menu.

3. Select Table Setup and select OK in the Create Table dialog box to accept the defaults.

4. Select Create Table, and save it with the name and location \Md2book\Drawings\Chapter5**Ex5-5**.

5. In Excel, fill in the spreadsheet as shown in Figure 5.18 and then save the file.

Figure 5.18

	A	B	C	D	E	F	G
1		Head_Dia	Head_Thick	Shaft_Dia	Shaft_Length	Hole_Dia	Hole_Depth
2	A	1	0.125	0.5	2	0.0625	0.5
3	B	1	0.125	0.5	3	0.0625	0.5
4	C	1.25	0.125	0.625	4	0.125	0.5
5	D	1.25	0.1875	0.625	5	0.125	0.5
6	E	1.5	0.1875	0.75	6	0.125	0.5
7	F	1.5	0.25	0.75	7	0.1875	0.625

6. Make AutoCAD the active application.

7. Select Update Link, select the file \Md2book\Drawings\Chapter5\Ex5-5.xls and then select Open.

8. The design variables will fill in the design variable dialog box. Select OK to exit the command.

9. Double-click on the different versions in the browser to change the part.

10. Make Excel the active application and highlight all the cells that contain information. Highlight the cells by selecting a cell in one of the corners and, while the mouse button is depressed, drag it to the opposite corner.

11. Copy this to the clipboard: either choose Copy from the Edit pull-down menu, hold down **Ctrl** and **C** at the same time or right-click and select Copy.

12. Make AutoCAD the active application and from the Edit pull-down menu, choose Paste Special and then select Paste Link from the dialog box. The Microsoft Excel worksheet should be highlighted. Then select OK and the chart will appear in the drawing. Your drawing should look like Figure 5.19.

13. Make Excel the active application by double-clicking in the linked chart and change a few of the cells in the spreadsheet. Then save the file.

14. Make AutoCAD active; the linked chart should automatically be updated. To update the variables, right-click on the name "Table (Ex5-5.xls)" in the browser and select Update.

Advanced Dimensioning, Constraining and Sketching Techniques

Figure 5.19

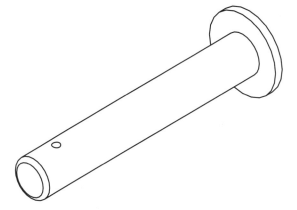

15. Double-click on the different versions in the browser to change the part.
16. Update the part and save the file.

Close Edge

In Chapter 1, you learned that the first sketch needed to be closed. In this section, you will learn how to create sketches that are opened and to close them by using existing edges of the part. If necessary, an open sketch can use more than one edge to close the sketch during profiling. When using the close edge option, you will follow the same steps as you previously learned: make a plane the active sketch plane and draw a sketch. If the sketch can use an existing edge to close the profile, do not draw that side. The open geometry cannot have a gap larger than the pickbox size. After you profile the sketch of an open profile, a message will appear:

```
Select edge to close profile:
```

Select the edge(s) to close the profile and press [⏎]. Silhouette edges cannot be used for closing the sketch. The number of constraints and dimensions that the profile needs will appear on the command line. Constrain and dimension the sketch as needed.

Tutorial 5.6—Closing a Profile by Selecting Edges

1. Open the file \Md2book\Drawings\Chapter5\Ex5-6.dwg.
2. Make the top horizontal plane the active sketch plane.
3. Draw in the two lines as shown in Figure 5.20.

Figure 5.20

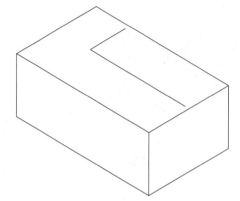

4. Profile the two lines and press [⏎] and when prompted:

   ```
   Select edge to close profile:
   ```

   select the top back horizontal and the top right vertical edges and press [⏎]. If you get the same message, stretch the geometry closer to the edges, zoom out from the part or increase the pickbox size and try it again.

5. Dimension the profile with a "1" and "1.5" value as shown in Figure 5.21.

Figure 5.21

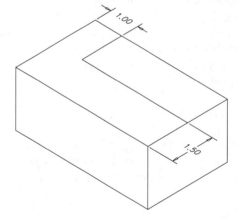

6. Extrude the profile ".75" with the Join option and accept the default direction. When complete, your part should look like Figure 5.22.

7. Make the right vertical plane the active sketch plane.

8. Draw in the three lines as shown in Figure 5.23.

Figure 5.22

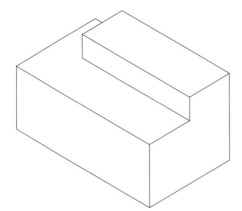

Figure 5.23

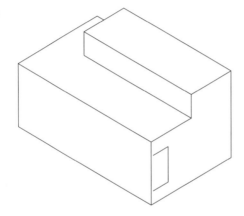

9. Profile the three lines and press [Enter]. When prompted:

   Select edge to close profile:

   select the front right vertical edge and press [Enter].

10. Dimension the profile with three ".5" dimensions as shown in Figure 5.24.

11. With the Cut option, extrude the profile through the part. When complete, your part should look like Figure 5.25.

12. Save the file.

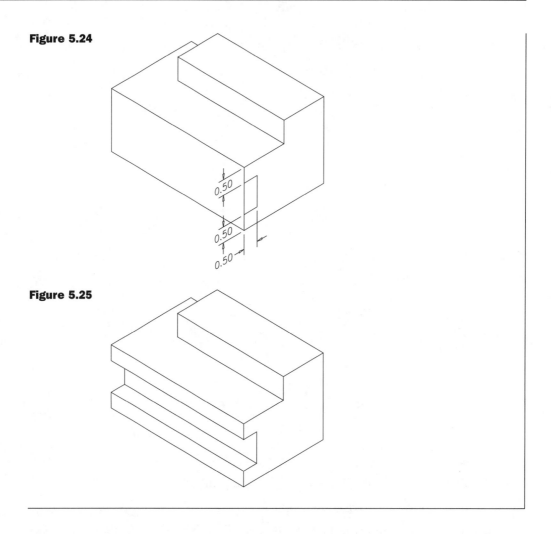

Figure 5.24

Figure 5.25

Copy Sketch

With the Copy Sketch command (AMCOPYSKETCH) as shown in Figure 5.26 from the Part Modeling toolbar, you can copy a sketch or a sketch of an existing feature to the active sketch plane of the active part. Sketches can be copied from one part to another. The sketch will be copied to the current sketch plane, along with the dimensions that define its shape. The dimensions that determine the sketch's X and Y placement will not be copied. Once on this plane you can rotate or move the sketch with AutoCAD commands and add dimensions to lock the sketch down to the current sketch plane. You can have multiple sketches on the part at the same time. When multiple sketches exist and you begin to create a feature, you will be prompted to select the sketch that you want. Access the Copy Sketch command through the Part Modeling toolbar, as shown

Advanced Dimensioning, Constraining and Sketching Techniques

in Figure 5.26, or right-click on the sketch name in the browser and then choose Copy from the pop-up menu.

Figure 5.26

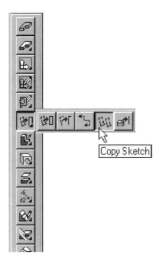

Tutorial 5.7—Copying a Sketch

In this tutorial there is a triangle that has been profiled, dimensioned and extruded to remove material, and a circle has been profiled and dimensioned.

1. Open the file \Md2book\Drawings\Chapter5\Ex5-7.dwg.

2. Issue the Copy Sketch command (AMCOPYSKETCH) or right-click on the name "SKETCH3" in the browser and choose Copy from the pop-up menu.

3. Press [↵Enter] to copy a sketch for the location. Select a point near the middle right of the top plane and press [↵Enter] to accept this location. Your drawing should look like Figure 5.27. At this point, you could dimension the circle to give X and Y placement on the top plane.

4. Make the angled face the active sketch plane and orient the positive X so that it is going into the screen and the positive Y is going toward the top of the part.

5. Issue the Copy Sketch command (AMCOPYSKETCH).

6. Press F and [↵Enter] to copy a sketch of a feature. Select the extrusion of the triangle and then select a point near the middle of the angled plane and press [↵Enter] to accept this location. Your drawing should look like Figure 5.28.

Figure 5.27

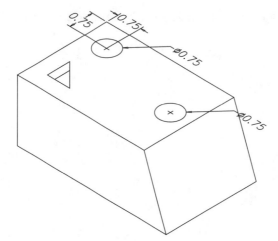

Figure 5.28

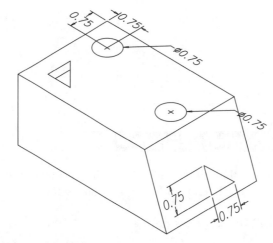

7. Add a ".5" and a "1.25" dimension to the triangle as shown in Figure 5.29.

8. Issue the Extrude command (AMEXTRUDE) and select Join, Blind, type a distance of ".5" and then select OK.

9. Select the back circle for the sketch to extrude and press [←Enter] to accept the default direction.

10. Press [←Enter] to repeat the Extrude command and pick OK to accept the default entries.

11. Select the front triangle for the sketch to extrude and [←Enter] to accept the default direction.

12. Press [←Enter] to repeat the Extrude command and select Cut for the Operation and then pick OK to accept the default entries.

Advanced Dimensioning, Constraining and Sketching Techniques

Figure 5.29

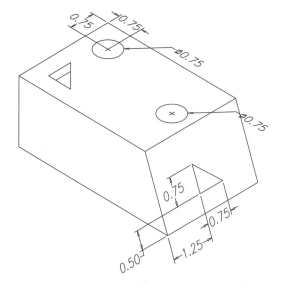

13. Since this is the only sketch left you can [←Enter] to accept the default direction. Your part should look like Figure 5.30.

Figure 5.30

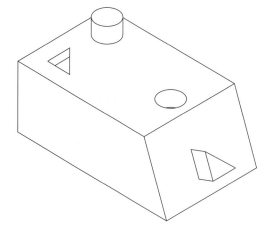

14. Save the file.

Construction Geometry

Construction geometry can help you create part that without them would be difficult to create. In chapter one you learned that sketches that are created in a continuous linetype, will be made into the part, any other linetype will be considered construction geometry. Create the sketch with construction geometry, then profile the sketch along with the construction geometry. Construction geometry can be constrained and dimensioned like continuous lines, but they will not seen in the part. Construction geometry can reduce the number of constraints and dimensions that are required to fully constrain a sketch. They can also be used to dimension to quadrants of arcs and circles. Construction geometry can be used to help define the sketch. For example, a construction circle inside a hexagon could drive the size of the hexagon. Without construction geometry, the hexagon would require six constraints and dimensions; it would require only three constraints and dimensions with construction geometry—the circle will have tangent constraints applied to it. When you edit a feature that was created with construction geometry, the construction geometry will reappear for the editing and disappear again when the part is updated.

Tutorial 5.8—Using Construction Geometry in Creating a Hexagon

1. Open the file \Md2book\Drawings\Chapter5\Ex5-8.dwg.
2. Change the circle to any linetype other than continuous.
3. Profile both the hexagon and the circle.
4. Show all your constraints; there should be six tangent and two sets of parallel constraints.
5. Add two angle dimensions and a diameter dimension as shown in Figure 5.31.

Figure 5.31

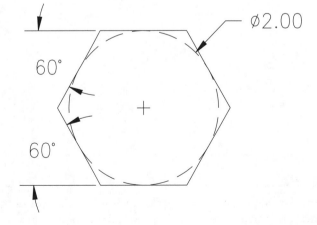

6. Use the Change Dimension command (AMMODDIM) to change the diameter of the circle to 1.

Advanced Dimensioning, Constraining and Sketching Techniques

7. Change to an isometric view (the **8** key).
8. Extrude the profile .5" in the positive Z direction.
9. Edit the part, change the diameter to 3 and update the part.
10. Save the file.
11. Start a new drawing, draw a hexagon and try to constrain it without using construction geometry. Then try to change the size of the hexagon.

Tutorial 5.9—Using Construction Geometry to Dimension to a Quadrant

1. Open the file \Md2book\Drawings\Chapter5\Ex5-9.dwg.
2. Change the top line to any linetype other than continuous.
3. Profile the construction line and the continuous lines and arc. There should be five dimensions or constraints required.
4. Show all your constraints; there should be two horizontal and two vertical constraints. There will not be a tangent constraint between the construction line and the arc.
5. Add a tangent constraint between the construction line and the arc.
6. Add dimensions as shown in Figure 5.32. The zero dimension is between the top of the two vertical lines. If you select the arc, it will go to the center of the arc. The zero dimension holds the arc in the middle of the geometry, no matter what the horizontal dimension is.

Figure 5.32

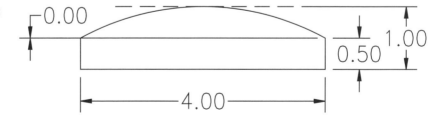

7. Change the "1" dimension to ".75".
8. Erase the ".5" dimension and create a "10" radial dimension in its place.
9. Change the "10" radial dimension to "15".

10. Extrude the sketch "4".

11. Save the file.

12. Start a new drawing and try to create the same part without construction geometry. Try to control the overall height of the part.

Converting Existing 2D Drawings to 3D Parametric Parts

A common question: How do I convert my existing 2D drawings to 3D parametric parts? Even though you cannot automatically convert the entire drawing, you can take the main shape with its dimensions, profile it and turn it into a part. Open the 2D AutoCAD drawing and either WBLOCK or copy to the clipboard the shape with its dimensions that best defines the shape. The copied geometry needs to follow the same rules as those of sketches: no large gaps or overlaps, and the dimensions need to have been placed to the endpoints and center of the geometry. Then start a new drawing and insert or paste the geometry into this drawing; explode the block. Then profile the geometry, along with all the dimensions. If you get an error message:

```
Highlighted dimension could not be attached
```

this means that these dimensions have not been created correctly or would form an over-constrained sketch. Erase the highlighted dimension and re-profile the sketch and dimensions. The dimensions that were profiled will automatically become parametric dimensions. Add any missing dimensions or constraints and then create a part. To add other profiles from an existing 2D drawing, make the plane the active sketch plane (where the profile will be placed) and then follow the same procedure.

Note: When copying geometry from a 2D drawing, do not copy islands. They will need to be created after the part is generated.

Advanced Dimensioning, Constraining and Sketching Techniques

Tutorial 5.10—Converting a 2D Drawing to a 3D Part

1. Open the file `\Md2book\Drawings\Chapter5\Ex5-10.dwg`.
2. Copy to the clipboard (**Ctrl+C** or WBLOCK) the outside profile of the top view (drawn as polyline) along with the dimensions, as shown in Figure 5.33.

Figure 5.33

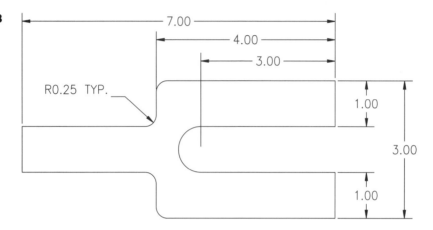

3. Start a new drawing.
4. Paste the geometry and dimensions from the clipboard, using **Ctrl+V** or insert them from the file you WBLOCKed out.
5. Explode the block with the EXPLODE command.
6. Profile the geometry and the dimensions. The sketch should be fully constrained and the dimensions should be parametric dimensions.
7. Change a few dimensions with the Change Dimension command (AMMODDIM) to prove that they are now parametric dimensions.
8. Change to an isometric view (the **8** key).
9. Extrude the profile "1".
10. Add the remaining features, holes and chamfers, as shown in Figure 5.34.
11. Save the file.

Figure 5.34

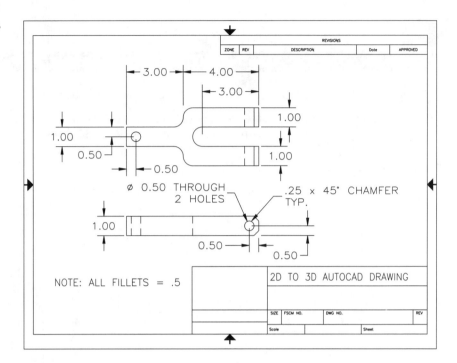

 Exercises

For the following three exercises, follow the instructions before each exercise.

Exercise 5.1—Vice Base

From scratch, create the part as shown in Figure 5.35. When creating this part, try using close edge, copy feature, and extrude To Plane and From Hole. For clarity, the part is shown with no fillets. The 3/4" ACME thread has a diameter of .58 and a tap diameter of .77. When the part is finished, try to create different types of fillets. Save the file as \Md2book\Drawings\Chapter5\ViceBase.dwg.

Advanced Dimensioning, Constraining and Sketching Techniques

Figure 5.35

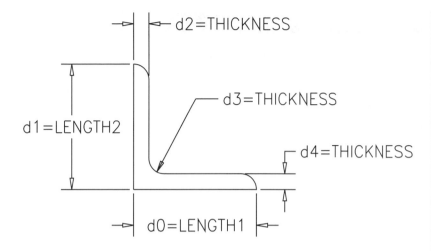

Exercise 5.2—Table-driven Angle

In this exercise, draw an angle, create table-driven variables and dimension the part as shown in Figure 5.36. Then link the Excel cells to the AutoCAD drawing. Save the file as \Md2book\Drawings\Chapter5**Angle.dwg**.

Figure 5.36

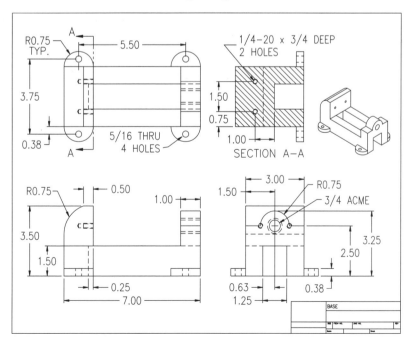

Exercise 5.3—Piston

In this exercise, open the file \Md2book\Drawings\Chapter5\Piston2D and turn it into a 3D parametric part. After creating the part, create the 1 1/2 diameter through hole. When the part is complete, save it as \Md2book\Drawings\Chapter5\Piston3D.dwg.

Figure 5.37

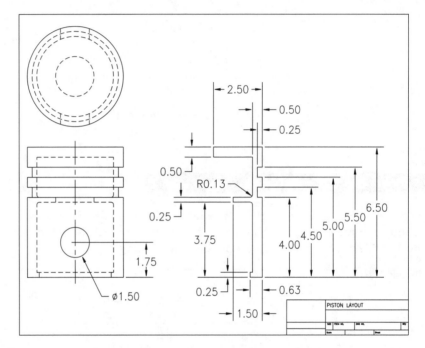

Advanced Dimensioning, Constraining and Sketching Techniques

Figure 5.38

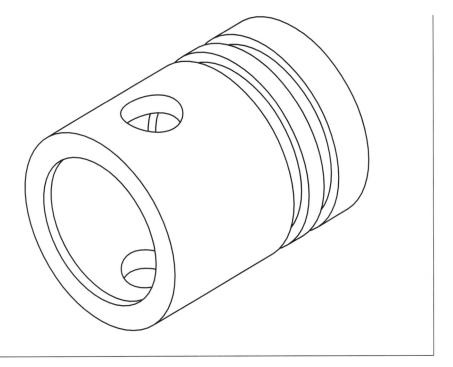

Review Questions

1. After a parametric dimension is erased, the next parametric dimension created will use the "d#" of the erased dimension. T or F?

2. When creating design variables, you can use any alpha numeric or numeric alpha combination you want. T or F?

3. List five numeric operators that can be used in equations.

4. Table-driven variables can be created with Lotus. T or F?

5. What is the difference between an active part variable and a global design variable?

6. An active part variable takes precedence over a global variable. T or F?

7. Silhouette edges can be used to close an open sketch. T or F?

8. When would you use construction geometry?

9. Only hidden lines can be used as construction geometry. T or F?

10. When a sketch is created from an existing 2D drawing, the dimensions cannot be included, but need to be recreated. T of F?

chapter 6

Advanced Modeling Techniques

Up to now, you have learned about basic modeling techniques. In this chapter, you will learn more advanced modeling techniques for creating multiple parts in the same file, mirroring a part, scaling a part, creating a thin-walled part using a shelling technique, copying features and combining two parts in a single part (using a technique called parametric booleans). You will also learn how to replay the steps in the creation of a part and get mass property information from a part.

After completing this chapter, you will be able to:

- Create multiple parts in a single file.
- Create and understand an "instance".
- Mirror a part.
- Scale a part.
- Create a shelled part.
- Copy features.
- Reorder features.
- Combine two parts in one part.
- Edit combined parts.
- Replay the steps in the creation of a part.
- Obtain mass property information from a part.

Adding a Part to a Drawing

As you work on a file, you may want to create multiple parts in the same file or work in what is known as assembly mode. There are three methods that can be used to create a new part in a drawing.

The first method is by using the browser. Click the right mouse button in a blank area in the browser. A pop-up menu will appear for selecting New Part. At the command line, you will see a prompt:

```
Select (or) <PART#>:
```

You can then type in a name for the part or press [⏎] to accept the default name. The default naming prefix can be set through Desktop preferences. By default, the naming prefix is set to Part. The first part created will have a name Part1_1, and each part created afterwards will increment by one, Part2_1, Part3_1. After each part has been created, its name will appear in the browser. The first number in the suffix represents the order in which the part was created. The second number refers to the instance number of a part. An instance is like an AutoCAD block. When a part is copied in the same file, it is created as an instance of the original part. The AutoCAD copy command will create an instance. If one of the instances changes, so will the other instances. (Instances will be covered in more detail in the next section.)

After the part has been given a name, create a sketch and profile it, as you have learned in the previous chapters. The Select option from the pop-up menu is used to convert a 3D solid (AutoCAD native solid) to a part that Mechanical Desktop can use. The part will not be parametric. However, features created to this part after the conversion will be parametric. Select a 3D solid and give it a name or press [⏎] to accept the default name.

The second method is to issue the New Part command (AMNEW) from the Part Modeling toolbar. After issuing the command, you will be prompted for the same information as in the browser method.

The third method is to make a new part based on an existing part, but this new part will not be an instance of the original part; instead it will be a new part definition. To create another part that is based on an existing part, use the Part Catalog command (AMCATALOG) found on the bottom of the browser or right-click on a part's name in the browser and select Copy. A dialog box will appear; select the All tab and all the part names will appear. Select Copy Definition; type in a new name and you will be returned to the drawing where you can insert this new part. This new part has no relationship to the original part. The Part Catalog command is discussed in greater detail in Chapter 7.

Instances

An instance is a copy of an existing part. An instance has the same name as the original part, but the number after the underscore will be sequenced. For example, if the original part has a name PART4_1, the copy will be PART4_2. An instance acts like a block in AutoCAD. If the original part or an instance of the part changes, all the parts will reflect the change. To create an instance, you can use the AutoCAD copy command, right-click on the part name in the browser and select Copy, or use the Part Catalog command

Advanced Modeling Techniques

Figure 6.1

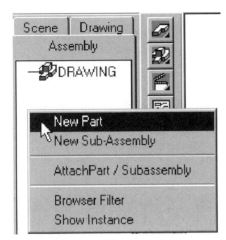

Figure 6.2

Figure 6.3

(AMCATALOG) with the Instance option. If you want a copy of the original part to have no relationship to the original part, use the Copy Definition option in the Part Catalog command.

Active Part

If there are multiple parts in the same file, only the active part can be edited. If you try to perform an operation on a part that is not active, an error message will appear at the command line:

```
Invalid selection. Keep trying.
```

Only one part can be active at a time. Looking at the Desktop browser, you can tell what part is active by the icon next to the name—if it is gray, it is the active part. To make a part active, you can double-click on the file name in the browser, select the part name in the browser with the right mouse button and select Activate Part, or issue the Activate Part command (AMACTIVATE) from the Part Modeling toolbar and select a part to make active.

Figure 6.4

Tutorial 6.1—Creating a New Part and Activating a Part

1. Open the file \Md2book\Drawings\Chapter6\Ex6-1.dwg.
2. Select in the white area of the browser with the right mouse button.
3. Choose the New part option from the pop-up menu.
4. Press to accept the default name.
5. Draw a circle.

Advanced Modeling Techniques

6. Profile the circle; it should require one constraint or dimension. This is because Part2_1 has no relationship to Part1_1.
7. Add a "2" diameter dimension to the circle.
8. Extrude the circle 2 units in the default direction.
9. Activate Part1_1 by right-clicking on the name Part1_1 and choose Activate Part from the menu.
10. Try to make the top plane of the circle that you just extruded the active sketch plane.
11. You will see the message `Invalid selection`.
12. Activate Part2_1 by right-clicking on the name Part2_1 and choose Activate Part.
13. Make the top plane of the circle the active sketch plane.
14. Issue the Part Catalog by selecting its icon from the bottom of the browser, go to the All tab, and copy part2 to Test2. Insert Test2 in the drawing by selecting an insertion point and press [⏎] to exit the command.
15. Edit the diameter of the part Test2_1 to ".5" and update the part. The Part2_1 should not have changed.
16. Use the AutoCAD copy command and copy part Test2_1 to a blank area on the screen. Then make this copy, Test2_2, the active part by double-clicking on the name Test2_2 in the browser.
17. Edit the diameter of the part Test2_2 to ".2" and update the part. The copy of the Test2_1 part should have also changed.
18. Save the file.

Mirroring a Part

After creating a part, you may need to create a second part that is a mirrored image of the first. A part can be mirrored about a planar face on the part, a work plane or a line. A line can be a regular AutoCAD line or two digitized points. When using the line method, you can use object snaps. Issue the Mirror Part command (AMMIRROR) from the Part Modeling toolbar, as shown in Figure 6.5, and at the command line you will see the prompt:

```
Select part to mirror:
```

Select the part to mirror (do not press [↵] after selecting the part, this would exit the command) and then you will be prompted:

```
Mirror about: Line/<Select planar face>:
```

If you want to select a plane, simply select in the middle of the planar face or select an edge that is part of the planar face. You will have an option either to accept the highlighted plane or to go to the next plane. When the correct planar face is highlighted, press [↵]. The next prompt will ask you either to:

```
Replace instances/<Create new part>:
```

If you press [↵], you will be prompted to give the new part a name. You can either accept the default or type in a different name. The part that was mirrored will be left alone and a new part will be created. This new part will have all the same dimensions and constraints of the original part, but it will not have a relationship to the first part (unless it was created with global design variables). If the variables are active part variables, the variables will be copied to the new part and have no relationship to the original part. If a global variable is used, the new part will maintain the relationship to the global variable. If you select the Replace option, the original part will be replaced and the mirrored part will take its place. To use the line option, press L and [↵]. Then you will be prompted to select two points. The two points must lie in the same plane; you can use object snaps.

Figure 6.5

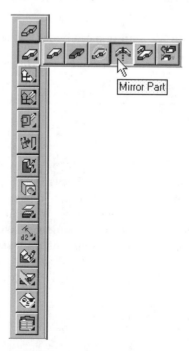

Tutorial 6.2—Mirroring About A Plane

1. Open the file \Md2book\Drawings\Chapter6\Ex6-2.dwg.

2. Issue the Mirror Part command (AMMIRROR).

3. Select the part on an edge but do not press [↲Enter], because this will exit the command.

4. Select the face as shown in Figure 6.6. Once it is highlighted, press [↲Enter] to accept this face.

Figure 6.6

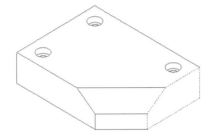

5. Press [↲Enter] to create a new part.

6. Press [↲Enter] to accept the default part name. The drawing should look like Figure 6.7.

Figure 6.7

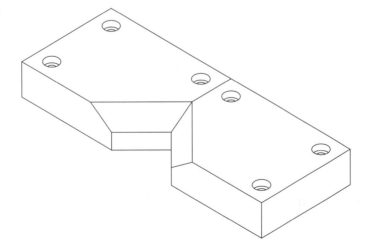

7. Save the file.

Mechanical Desktop 2.0: Applying Designer and Assembly Modules

Tutorial 6.3—Mirroring About a Line and Replacing the Part

1. Open the file \Md2book\Drawings\Chapter6\Ex6-3.dwg.

2. Issue the Mirror Part command (AMMIRROR).

3. Select the part on an edge but do not press [←Enter], because this will exit the command.

4. Press **L** and [←Enter] to use the Line option.

5. Use object snaps to select the endpoints of the line.

6. Press **R** and [←Enter] to Replace instances. When complete, your drawing should look like Figure 6.8.

Figure 6.8

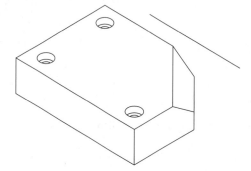

Scaling a Part

When creating parts, you may want to scale a part up or down or convert it from inches to mm or vice versa. The Scale Part command (AMSCALE) will scale a Mechanical Desktop part and maintain full parametrics. Issue the Scale Part command from the Part Modeling toolbar and select the part or parts that you want to scale; use any AutoCAD selection technique (pick, window, crossing etc.). When done, press [←Enter], then select a base point—you can use object snaps. The base point will be the point from which the part is scaled up or down. Then type in the scale factor to scale the part. A value greater than 1 will increase the size and any number less than 1 will decrease the part size. Each dimension will be scaled according to the scale factor. For example, a factor of 2 will double the size of the part, while .25 will decrease the part to one quarter of the original size. Angle dimensions will not be scaled. For example, a part with a 45° angle will still be 45° after scaling. Design variables will be converted to actual numbers. For example, if a design variable of length = 5 is scaled by the factor of .5, the result would be a "2.5" dimension. Under-constrained sketches can also be scaled. Table-driven parts cannot be scaled.

Figure 6.9

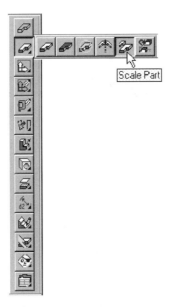

 Tip: If you convert many drawings between inches and millimeters, you can set up a global variable as a multiplier and use it everywhere you need a dimension. For example, set a global variable as mm=1. Then when you create the part, use mm as a multiplier for every number in the part. For example, for a "4" inch dimension that will also be converted to millimeters, type in "4*mm". Then when you want to convert the file to millimeters, change the global variable mm from 1 to 25.4 and the entire model will be converted to millimeters.

Tutorial 6.4—Scaling

1. Open the file \Md2book\Drawings\Chapter6\Ex6-4.dwg.
2. Issue the Edit Feature command (AMEDITFEAT) to verify the length of the part is "6.50".
3. Issue the Scale Part command (AMSCALE).
4. Select the part and press ⏎.
5. Select the center of the front circular edge.

6. Type in a value of ".5"and press [Enter].

7. Issue the Edit Feature command to verify the length of the part is "3.25".

8. Save the file.

Tutorial 6.5—Scaling a Part With Design Variables

1. Open the file \Md2book\Drawings\Chapter6\Ex6-5.dwg.

2. Verify that the design variables in this drawing are Length=4 and Width = 3.

3. Issue the Edit Feature command (AMEDITFEAT) to verify that the dimensions are driven by design variables.

4. Issue the Scale Part command (AMSCALE).

5. Select the part and press [Enter].

6. Select a point near the center of the part.

7. Type in a value of ".5"and press [Enter].

8. Issue the Edit Feature command to verify the design variables have been replaced with a "2" and "1.50" dimension.

9. Save the file.

Shelling

Shelling refers to giving the outside shape of a part a thickness (wall thickness) and removing the remaining material on the inside, like scooping out the inside of a part, leaving the walls a specified thickness. A part can only be shelled once, but individual faces of the part can have different thicknesses. This is referred to as a thickness override. If a face that you select for a thickness override has faces that are tangent to it, those faces will also have the same thickness override. Also, a face can be excluded from being shelled, which would leave it open. Once a part has been created that you want to shell, issue the Shell command (AMSHELL) from the Part Modeling toolbar and a dialog box will appear, as shown in Figure 6.11. The dialog box contains the following sections: Default Thickness, Excluded Faces and Multiple Thickness Overrides.

Default Thickness

Inside: Offsets the wall thickness by the given value into the part.

Advanced Modeling Techniques

Figure 6.10

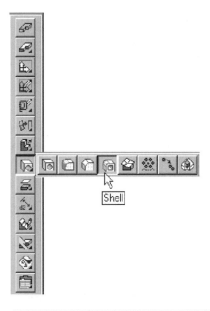

Figure 6.11

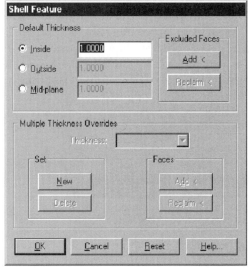

Outside: Offsets the wall thickness by the given value out of the part.

Mid-plane: Offsets the wall thickness evenly into and out of the part by the given value. A mid-plane offset cannot have overrides.

Excluded Faces

Add: The selected face will be left open.

Reclaim: The selected open face will be shelled or remained closed.

Multiple Thickness Overrides

Thickness: The override value of the offset. You can have multiple overrides for a single part applied to different faces.

Set

New: This is where you set the thickness override, then use Add to apply it to the selected face.

Delete: The selected overridden face will be set back the default thickness.

Faces

Add: The selected face(s) will be given the override value for the thickness.

Reclaim: The faces that were overridden with the current override will be highlighted. Then selecting any of the highlighted faces will return them to the default thickness.

After filling in the dialog box, select OK and the part will be shelled. To edit a shell, issue the Edit Feature command (AMEDITFEAT) or double-click on the shell in the browser. When you edit a shell, the same dialog box will appear as when the shell was created. Here you can change the default thickness and add or reclaim faces, as well as add or delete any of the thickness overrides that have been set. When editing a shell, follow the same steps as you would to create the shell.

Tutorial 6.6—Shelling and Edit Shell

1. Open the file \Md2book\Drawings\Chapter6\Ex6-6.dwg.

2. Issue the Shell command (AMSHELL).

3. Type in a value of ".03" for the Inside value and select OK. The inside of the box should be totally shelled out and look like Figure 6.12.

4. Edit the shell with the Edit Feature command (AMEDITFEAT) or from the browser, double-click on Shell1. From the dialog box select Add from the Excluded Faces section and select the top face, as shown in Figure 6.13.

5. In the dialog box, change the default thickness to ".1" and then select OK.

6. If the part did not automatically update, update it now. Your part should look like Figure 6.14, shown with lines hidden.

Figure 6.12

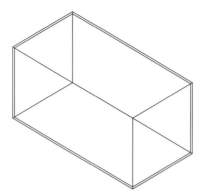

Figure 6.13

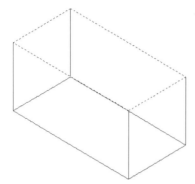

Figure 6.14

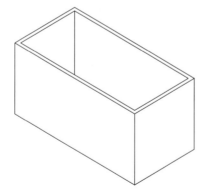

7. Edit the shell, select New from the Multiple Thickness Overrides section and type in a thickness of ".2".

8. Select the Add button under Faces, select the front left face, as shown in Figure 6.15, and press ⏎ twice to return to the dialog box.

9. Select OK and if the part did not automatically update, update it now. Your part should look like Figure 6.16, shown with lines hidden.

Figure 6.15

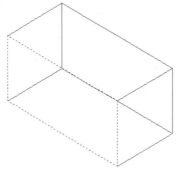

Figure 6.16

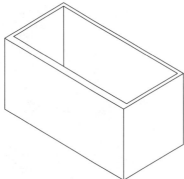

10. Edit the shell and Select Reclaim from the Faces section near the bottom right of the dialog box.

11. Select the same face as you did in step 8.

12. Back in the dialog box, select OK and update the part if necessary. Your part should again look like Figure 6.14.

13. Save the file.

Tutorial 6.7—Shelling

1. Open the file \Md2book\Drawings\Chapter6\Ex6-7.dwg.

2. Issue the Shell command (AMSHELL).

3. Give a value of ".06" for an Outside value.

4. Select Add from the Excluded Faces section and select the top and bottom faces of the revolution, as shown in Figure 6.17. Select the top face and press [⏎], then select the bottom face. You may need to advance to the next face if it is not highlighted. Press [⏎] twice to return to the dialog box.

5. Select New from the Multiple Thickness Overrides section and type in a value of ".125".

Figure 6.17

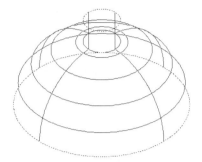

6. Select the Add button under the Faces section and select the top cylinder as shown in Figure 6.18. Press [⏎ Enter] twice to return to the dialog box and then select OK in the dialog box.

7. Update the part if necessary. When complete, your part should look like Figure 6.19.

Figure 6.18

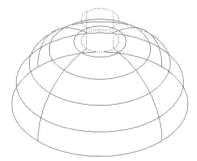

Figure 6.19

8. Save the file.

Copying a Feature

In the previous chapter, you learned that you can copy a sketch and then create a feature based on that sketch. In this section, you will learn how to copy a feature to the current sketch plane. Features can be copied to the same part or to a different part. However, the parts need to be in the same file and the part that the feature is being copied to must be the active part. The following features cannot be copied: the base feature, fillets, chamfers, shells, features within a toolbody (parametric Boolean), features where the sketch was closed by an edge and features that have a termination To Plane, To Face and From/To.

Before copying a feature, make the target part active. Then make the plane where the feature will be placed the active sketch plane. Issue the Copy Feature command (AMCOPYFEAT) either by right-clicking on the feature's name in the browser and choosing Copy from the pop-up menu or by selecting the Copy Feature command from the Part Modeling toolbar, as shown in Figure 6.20. After issuing the command from the toolbar, select the feature to copied in the drawing screen. Depending on the feature that was selected, you may be prompted to accept or cycle to the next feature. If the command was selected from the browser, the feature will already have been selected. Select a point with the left mouse button where you want the feature to be placed on the active sketch plane. The copied feature will appear in red. To change the position, keep selecting points with the left mouse button.

Options

At the command line you may see the options:

`Parameters/Rotate/Flip/<Select location>:`

Not all of the options may appear, depending on the geometry that is being copied.

Parameters: Allows you to set the copied feature as a dependent or independent feature of the original feature. Dependent means that if the parent dimension changes, so will the copied features. A dependent feature will have its dimensions d# equal to the d# of the original dimension. You can break this relationship by editing the feature and typing in a new value for any dimension. Now it will be independent, with no relationship to the original feature.

Rotate: Rotates the copied feature in 90° increments until it is in the correct orientation.

Flip: Mirrors the copied feature about the ZX plane of the current UCS.

After the feature is copied, it will not be constrained to the plane or part. If the feature needs to be constrained to the plane, use the Edit Feature command (AMEDITFEAT) with the Sketch option and add dimensions or constraints as needed. Then update the part.

Advanced Modeling Techniques

For holes that are copied, a work point is used to locate it, hence a work point will appear when you return to the sketch. Add dimension or constraints to the work point and update the part.

Figure 6.20

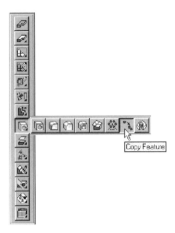

Tutorial 6.8—Copying a Feature

1. Open the file \Md2book\Drawings\Chapter6\Ex6-8.dwg.

2. Make the angled face the active sketch plane and orient the positive X so it goes into the screen.

3. Issue the Copy Feature command (AMCOPYFEAT) and select the extruded feature or right-click on the name ExtrusionBlind2 in the browser, choose Copy from the pop-up menu and then press [↵]. Select a point near the middle of the angled plane but do not press [↵].

4. Press **P** and [↵]; press **D** and [↵] to set up a dependency between the two features.

5. Press **R** and [↵] to rotate the feature so that it looks like Figure 6.21, and press [↵] to exit the command.

6. Edit the sketch of the copied feature and add both a ".5" and a "1" dimension, as shown in Figure 6.22.

7. Update the part.

8. Edit a few of the dimensions in the original extrusion and update the part. Both extrusions should look the same, since a dependency was set when the feature was copied.

Figure 6.21

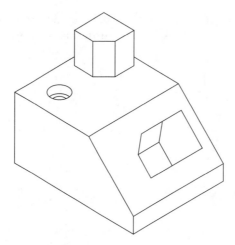

Figure 6.22

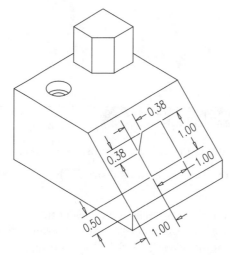

9. Save the file.
10. Type in "USC" and press [↵] twice to return to the world coordinate system.
11. Create a new part by clicking the right mouse button in the white area in the browser and selecting New Part. Press [↵] to accept the default name.
12. Draw a circle, profile it, give it a "2" diameter dimension and extrude it "2".
13. Make the top of the cylinder the active sketch plane.
14. Issue the Copy Feature command (AMCOPYFEAT) and select the hole feature on Part1_1. Press [↵] accept this feature, and select a point near the center of the top of the cylinder and press [↵].

15. Edit the sketch of the copied hole and add a concentric constraint between the work point and the top of the cylinder. Update the part; your drawing should look like Figure 6.23.

Figure 6.23

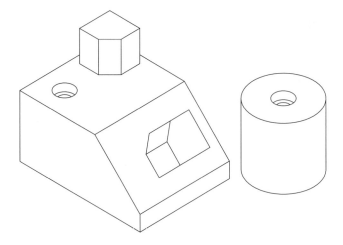

16. Save the file.

Reordering

As you build your part, you may get an unexpected result after a feature is created. After looking at the part, you may see that if a feature had been created in a different order, you might have achieved a different result. Instead of recreating the part, you can reorder the history of the features in the browser. Reordering can be done in the browser or through the Reorder Feature command (AMREORDFEAT). In the browser, with the left mouse button select a feature's name in the browser tree and, while keeping the mouse button depressed, move the feature up or down in the tree structure. As you move the feature through the tree structure, you will see a circle with a diagonal line through it. This means that the feature cannot be moved to this location. Keep moving the feature until a horizontal line appears in the browser. This tells you that the feature may be placed in this location. If this is the correct location, release the left mouse button. Otherwise, keep moving the feature until it is in the correct location. If you prefer not to use the browser, you can issue the Reorder Feature command (AMREORDFEAT) from the part Modeling toolbar and select the feature that you want to reorder. Then select the feature that you want to come before it.

Figure 6.24

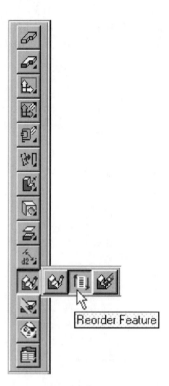

Tutorial 6.9—Reordering

1. Start a new drawing.
2. Create a 2" square and extrude it 1".
3. Change to an isometric view using the **8** key.
4. Create a .5" drill through hole 1" in from each edge from the bottom of the box.
5. Shell the cube with a thickness of .1 and exclude the top face (so that it is open). Your part should look like Figure 6.25. The drilled hole appears as a boss because the shell adds thickness to the hole.
6. If the result intended is a drilled through hole instead of a boss, it should have been created after the shell. In the browser, select the hole and, keeping the left mouse button depressed, move it so that it is after the shell, as shown in Figure 6.26. Then release the left mouse button. It did not matter if you reordered the hole after the shell or the shell before the hole. Your part should resemble Figure 6.27.

Advanced Modeling Techniques

Figure 6.25

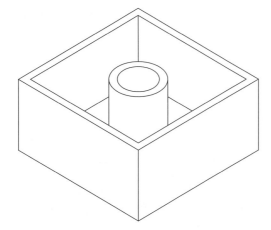

Figure 6.26

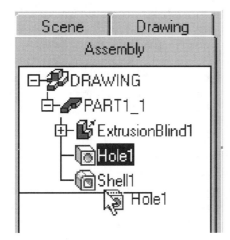

Figure 6.27

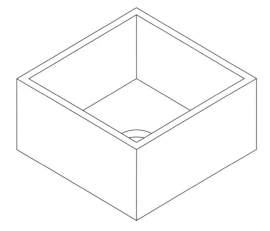

7. Move the shell so that it is after the hole and the part should again look like Figure 6.25.

8. Save the file as \Md2book\Drawings\Chapter6**Ex6-9.dwg**.

Creating Combined Parts

As you start creating parts, you may want to create a part that is made up of two individual parts. Instead of recreating this part, you can use the Combine command (AMCOMBINE), as shown in Figure 6.28, to cut, add or keep what is common between the two parts. For example, if you need to create a mold for a part that already exists, you could remove the part from a rectangular part. To create a part that consists of two different parts, start by creating two parts in the same file. Make the part that will be the base part active (the part that will have material added to it or removed from it). Then you can use AutoCAD commands to align the parts in the correct position or use assembly constraints (which will be covered in Chapter 7) to position the two parts in the correct location. If assembly constraints are used to locate the two parts, there will be a relationship between the parts. If one part changes shape, the second part will maintain the relationship that was set with assembly constraints. The two parts do not need to be touching; there can be space between them and they still can be joined together. Of course, there will be a gap even after they are joined.

Then issue Combine command (AMCOMBINE) from the Part Modeling toolbar. At the command line you will be prompted for which operation you want to perform: `<Cut>/Join/Intersect:` After selecting an operation, you will be asked to select a toolbody. A toolbody is consumed by the base part. After you select the toolbody, the command will be complete. In the browser, the toolbody will appear as a part nested under the name COMBINE#, which itself is nested under the active part.

Editing Combined Parts

The parametric dimensions for both parts are still intact, even though the parts have been combined. Once a combined part has been created, there are three methods for editing both parts, as explained below.

The first method is to use the Edit Feature command (AMEDITFEAT) and select any of the features. Once the feature is selected, its dimensions will appear on the screen. Select the specific dimension to edit, give it a new value and then update the part as you would any other part.

The second method also starts with the Edit Feature command, but instead of selecting the feature to edit, press T to edit the toolbody. This will place the part in a rolled back state, which means that the parts will appear as they were before the Combine command.

Advanced Modeling Techniques

Repeat the Edit Feature command to edit the toolbody features. The part that was the parent part cannot be edited in a rolled back state. After editing the toolbody, update the part. You will have three options at the command prompt: eXit/Full/<Active Part>:

eXit: Returns you to the rolled back state without updating the part.

Full: Updates the toolbody and returns the two parts in a combined state.

Active Part: Updates the toolbody only.

The third method is to select the part feature from the browser as you did in the previous section.

The steps in combining two parts:

1. Create two parts in the same drawing.

2. Make the main part the active part (the main part is the part to which geometry will be added to or from which geometry will be removed).

3. Position the parts with the AutoCAD commands like move, rotate etc. or use assembly constraints, which will be covered in Chapter 7.

4. Issue the Combine command (AMCOMBINE).

5. Select the Boolean operation Cut, Join or Intersect.

6. Select the part to be used as the toolbody.

Figure 6.28

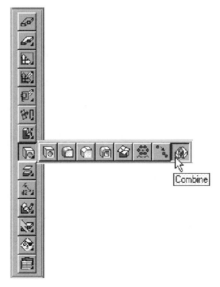

Mechanical Desktop 2.0: Applying Designer and Assembly Modules

Tutorial 6.10—Combining with Cut

In this example, the two parts are already in the correct position.

1. Open the file `\Md2book\Drawings\Chapter6\Ex6-10.dwg`.
2. Verify that the parts are individual parts by selecting each of the part names individually in the browser; they will highlight independently.
3. Rotate the parts again in shaded mode to show the cavity.
4. Make the part "MoldBase" the active part.
5. Issue the Combine command (AMCOMBINE), use the Cut option and select the mouse part.
6. Rotate the parts again in shaded mode to show the cavity.
7. Use the Edit Feature command (AMEDITFEAT) to edit the cavity of the mouse part, change the "1" depth dimension to "1.25" and update the part.
8. Save the file.

Tutorial 6.11—Combining with Join

1. Start a new drawing.
2. Create a new part named Shaft.
3. Draw, profile and dimension as shown in Figure 6.29.

Figure 6.29

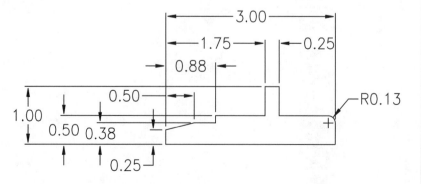

4. Do a full revolution along the bottom horizontal line.
5. Create a new part named Base.
6. Draw a 4" square and extrude it 2".
7. Move the center of the Shaft to the middle of the Base using object snaps and the move command. (Assembly constraints can also be used to position the parts.)
8. Issue the Combine command (AMCOMBINE), use the Join option and select the Shaft. When complete, your drawing should look like Figure 6.30.

Figure 6.30

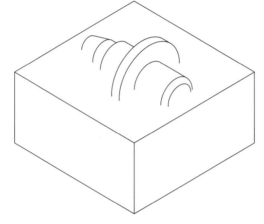

9. Use the Edit Feature command (AMEDITFEAT) and use the Toolbody option. This will take the Shaft back to its condition before the Combine command.
10. Issue the Edit Feature command again and select the shaft, change the overall length from "3" to"4" and update the part with the Full option. Your part should look like Figure 6.31.

Figure 6.31

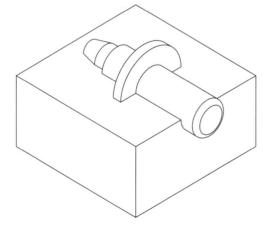

11. Edit the length of the Base from "4" to "5" (the horizontal dimension along the X axis).

12. Do a Full update. When complete, your part should resemble Figure 6.32. If assembly constraints were used to position the part, you could control the shaft's positioning in the base.

Figure 6.32

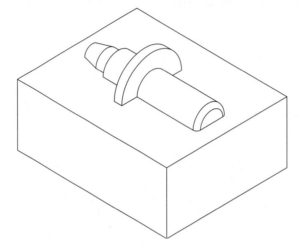

13. Save the file as \Md2book\Drawings\Chapter6\Ex6-11.dwg.

Replaying

In AutoCAD, there is the UNDO command that steps back one command at a time. Since Mechanical Desktop is running inside AutoCAD, the UNDO command is still a valid command. There is also in AutoCAD a REDO command that undoes an undo, but this will only undo the last undo. In Mechanical Desktop, there is a command called Feature Replay (AMREPLAY) that is similar to redo because it steps forward through the command sequence in which the part was created. The Feature Replay command can be used to see how a part was built, to step through a part and stop at a specific point and take a different approach. If you receive an error message, it can be used to replay through the part and truncate the part at the point before the error message appears. While stepping through the part, you will see at the command line the commands that were used. Also, as you step through the replay, the feature names will highlight in the browser in the order in which they were created.

After issuing the Feature Replay command from the Part Modeling toolbar, you will be prompted:

Advanced Modeling Techniques

```
Display/Size/Truncate/eXit/<Next>:
```

Display: Turns on all the constraint symbols. Once they are turned on, they cannot be turned off for the duration of the Feature Replay command.

Size: Changes the height of the constraints. This can be adjusted throughout the command.

Truncate: Stops the command. The part will be left in the state in which it was truncated.

Exit: Exits the command. The part will be returned to its state before the Feature Replay command.

Next: The last option steps to the next command used in building the part. After stepping all the way through the building process, you will be returned to the command prompt.

Notes: While in the Feature Replay command, you can transparently pan and zoom.

The Feature Replay command is not used to reorder the part. If you want to reorder the part, use the browser.

After a part is truncated, the features that were created after this point will be discarded from the part.

Tip: After receiving complex parts, use the Feature Replay command to gain insight into how the part was created. While running through the sequence, think about other methods that could have been used to create the part.

Figure 6.33

Tutorial 6.12—Replaying the Command Sequence

1. Open the file \Md2book\Drawings\Chapter6\Ex6-12.dwg.
2. Issue the Feature Replay command (AMREPLAY).
3. Press **D** and [⏎] to turn on all constraint symbols.
4. Press **S** and [⏎] to adjust the constraint symbols to a larger size and then select OK. If the size is not legible, go back and readjust the size.
5. Press [⏎] to cycle all the way through the part, watching the command line and the browser as you cycle through.
6. Issue the Feature Replay command (AMREPLAY) again or press [⏎] to repeat the Feature Replay command.
7. Cycle through the part and then truncate it after both holes are created. When complete, your part should resemble Figure 6.34.

Figure 6.34

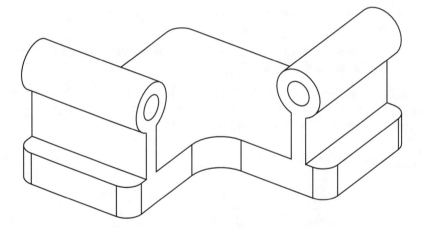

8. Add five separate ".125" diameter through holes that are concentric to the part, as shown in Figure 6.35.
9. Issue the Feature Replay command (AMREPLAY).
10. Cycle through the entire part.
11. Save the file.

Figure 6.35

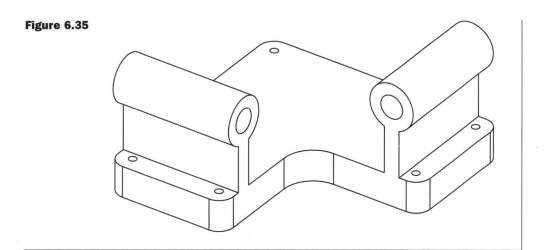

Mass Properties Information for the Active Part

To obtain mass property information about a part, first make it the active part. Then issue the Active Part Mass Properties command (AMPARTPROP) from the Part Modeling toolbar and a dialog box will appear. Type in the density of the material and press ⏎, and the mass property information will be updated. To create an ASCII file of this information, select the Write to File button and give the file a name and location.

Figure 6.36

Mass Properties for Multiple Parts

If you want mass property information for multiple parts or want more control of how the numbers are calculated, issue the Assembly Queries command (AMASSMPROP) from the Assembly Modeling toolbar. After issuing the command, you will be prompted to

```
Select parts/subassemblies Name/<Select>:
```

If you know the subassembly name, you can type it in or press [⏎] to select a single part or multiple parts. When selecting parts, you can use any AutoCAD selection technique. After making the selection, press [⏎] and a dialog box will appear containing five areas: % Error, Coordinate System, Material, Assembly Units and Mass Units.

% Error: The default is "1". The lower the number, the more accurate the calculations, but the longer the calculation time.

Coordinate System: The choices are:

  Parts CG (center of gravity). This can only be used for a single part.

  UCS, the current coordinate system.

  WCS, the world coordinate system.

Material: Selecting on Material will bring up a dialog box. On the right side of the dialog box select the part or parts that you want to assign a specific material. To select multiple parts, hold down the **Ctrl** key while selecting the part names. After selecting the part(s), select in the material area, or select the down arrow, and the available materials will appear. Select a material, then select Assign and the material will be assigned to the part(s). Continue until all parts have a material assigned to them. To add materials to the list, open the ASCII file named MCAD.mat in the AutoCAD R14\Desktop\support directory and add the property information. Use the SaveAs option to append it to the MCAD.mat file.

Assembly Units: Select the units that the parts were drawn in: In, cm or mm.

Mass Units: Select the units that you want the mass in: Lbs, g or Kg.

After filling in the information, select OK and the mass property information will appear in a dialog box. To write the information out to an ACII file, select File or to exit the command, select Done.

Advanced Modeling Techniques

Figure 6.37

Tip: To receive mass property information that is always based upon the same UCS orientation, use the world coordinate system and do not move the part.

Tutorial 6.13—Obtaining Mass Property Information

1. Open the file \Md2book\Drawings\Chapter6\Ex6-13.dwg.
2. Make the World coordinate system current, at the command line type in "UCS" and press ⏎ twice.
3. Issue the Active Part Mass Properties command (AMPARTPROP).
4. Type in a density of ".2" and press ⏎.
5. Write the information out to a text file: \Md2book\Drawings\Chapter6**Ex6-13A.ppr.**
6. Issue the Assembly Queries command (AMASSMPROP) and select both parts.

Mechanical Desktop 2.0: Applying Designer and Assembly Modules

7. Set the Coordinate System to WCS, Assembly units to In, Mass units to Lbs.
8. Assign Material for Part1 to Mild steel and Part2 to Aluminum.
9. To see the mass property information, select OK in both dialog boxes.
10. Select File and create an ASCII file: `\Md2book\Drawings\Chapter6\Ex6-13B.ppr`.
11. Save the file.

Exercises

Follow the instructions before each exercise.

Exercise 6.1—Bottom Half of Utility Knife

Create the bottom half of a utility knife, as shown in Figure 6.38. If you have trouble creating the part, you can open the finished part

`\Md2book\Drawings\Chapter6\Uknife1 done.dwg` and replay the sequence that was used to create the part. When done, save the file as `\Md2book\Drawings\Chapter6\Uknife1.dwg`.

Figure 6.38

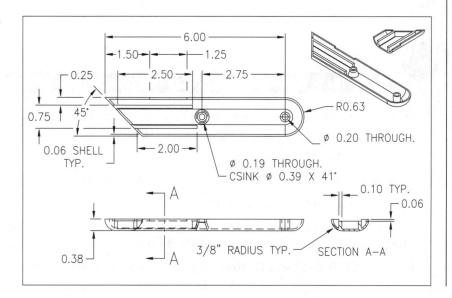

Exercise 6.2—Top Half of Utility Knife

Create the top half of a utility knife, as shown in Figure 6.39. Use the part created in Exercise 6.1 to mirror the bottom half of the utility knife. Again, if you have trouble creating the part, you can open the finished part

\Md2book\Drawings\Chapter6\Uknive2 done.dwg and replay the sequence that was used to create the part. When done, save the file as \Md2book\Drawings\Chapter6\Uknife2.dwg.

Figure 6.39

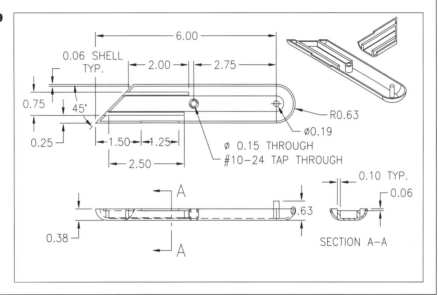

Review Questions

1. List the steps that are required to create a new part using the browser.

2. A feature can only be edited from the active part. T or F?

3. After a part is mirrored with the Create new part option, the new part has no relationship to the original, except if it was created with global variables. T or F?

4. What happens to the design variables in a part that is scaled?

5. When a part is shelled that has no tangent faces, the part can have multiple wall thicknesses. T or F?

6. In shelling, what does an excluded face refer to?

7. After two parts are combined, parametric dimensions are lost for both parts. T or F?

8. When editing a combined part, you can get into a rolled back state. What is a rolled back state? What is the option to get there?

9. The replay command can reorder features. T or F?

10. Mass property information is always based on the World Coordinate System. T of F?

11. What is a toolbody?

12. Can you copy a feature from one part to another?

13. How do you tie down a copied feature?

14. What is the difference between copying a part and copying a part's definition?

chapter 7

Assemblies

An assembly file is a file that contains more than a single part. The parts may be contained within the same file or they may be referenced in an assembly file.

In this chapter, you will learn how to create "top down" and "bottom up" assemblies, constrain the parts to one another using intelligent constraints, edit the constraints, check for interference and create exploded views. Mechanical Desktop 2.0 uses a variational solver for constraining parts, which means that the order of the parts in an assembly is not important, except that the base part (component) will be grounded and the other parts will move to it. This type of assembler is needed to assemble linkage types of mechanisms.

After completing this chapter, you will be able to:

- Create top down assemblies.
- Create bottom up assemblies.
- Constrain parts together using assembly constraints.
- Edit assembly constraints.
- Check for interference.
- Create scenes and exploded assembly views.
- Edit scenes and add trails.

Creating Assemblies

By default, Mechanical Desktop starts up in assembly mode, with the Assembly tab in the browser as the active tab. In assembly mode, you can create a single part or multiple parts.

If you want to create just a single part, start a new file under the file pull-down menu with the New Part File option. With that option, you cannot create more than a single part in that file, and assembly-related commands will be disabled.

There are three ways to create assemblies using Mechanical Desktop: top down, bottom up and assemblies that are a combination of both top down and bottom up techniques.

A top down approach refers to an assembly where all the parts are located in the same file. In other words, the user creates each part from within the top-level assembly. These parts can later be externalized and referenced to the assembly.

Bottom up refers to an assembly where all the parts are external to the assembly file and referenced in. The user creates the parts in their own files and then builds the assembly by referencing the parts to it.

Top Down Approach

In the last chapter, you created a top down assembly by creating multiple parts in the same file. In this section, you will create another top down assembly, and later in the chapter you will apply assembly constraints to position the parts. To create multiple parts in the same file, you can right-click in the white area of the browser. Select the New Part option or issue the New Part command (AMNEW) from the Part Modeling toolbar, then type in a name for the part or press to accept the default name.

Note: You must activate a part in order to edit a part.

Catalog

Issue the Part Catalog command (AMCATALOG) by clicking on the Catalog icon on the bottom of the browser as shown in Figure 7.1. Right-click in the white area of the browser and select Attach Definition or right-click on a part name in the browser and select Show Definition. An Assembly Catalog dialog box will appear, as shown in Figure 7.2. In this dialog box, you will see what parts are in the file, and you can insert an instance or reference in a file. The Assembly Catalog dialog box has two tabs: External and All. The External tab shows all the external files that are attached to the current assembly file. Details about this tab will be covered in the bottom up assembly section. The All tab shows all the external files that are attached to the file in the left area of the browser and all the local files in the assembly on the right side of the dialog box. If there are local parts in the assembly, you can right-click on the part name and choose one of the six options from the pop-up menu as shown in Figure 7.3.

Assemblies

Figure 7.1

Figure 7.2

Figure 7.3

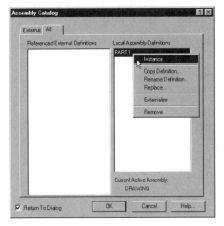

Instance: Creates an instance of the part, which is like an AutoCAD block. If one instance changes, so will the other instances.

Copy Definition: Creates a new part based on the selected part. This new part has no relationship to the original part.

Rename Definition: Gives the selected part a new name.

Mechanical Desktop 2.0: Applying Designer and Assembly Modules

Replace: Replaces the selected part with another part in the current file. The selected part will maintain its current name, but the geometry will be replaced by the new part.

Externalize: Writes the selected part out to a new file. This file is then linked or referenced to the current file.

Remove: Erases the part from the current file.

Tutorial 7.1—Top Down Assembly

1. Start a new file.
2. Draw, profile and dimension a "4" square and extrude it ".5" in the default direction.
3. Create a new part with the default name PART2.
4. Draw an "L" bracket as shown in Figure 7.4.

Figure 7.4

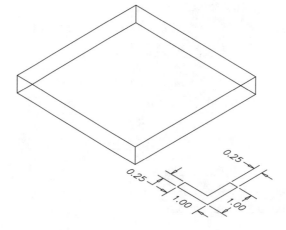

5. Extrude the "L" bracket ".25" in the default direction.
6. Create a new part with the default name PART3.
7. Draw, profile and dimension a ".5" diameter circle, extrude it "2". When complete, your drawing should look like Figure 7.5.
8. Right-click on the name Part2_1, select the copy option and select a point to the left of the current "L" bracket.

Figure 7.5

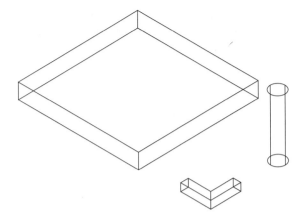

9. Issue the Part Catalog command (AMCATALOG).
10. Select the All tab and notice the names of the parts.
11. Right-click on the name Part2, select Copy Definition and give it a name "LFILLET" and then place 2 instances of the new part near the lower left part of the drawing.
12. Make "LFILLET_2" the active part.
13. Add a ".5" fillet to the inside edge of the bracket and update the part. You may want to switch your viewpoint to see the inside edge better. Your drawing should look like Figure 7.6.

Figure 7.6

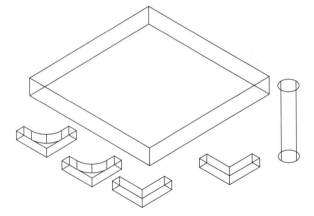

14. Issue the Part Catalog command (AMCATALOG), then right-click on Part3 and select the Remove option. Select OK to confirm that this part will be deleted from the file.

15. Issue the Part Catalog command, then right-click on Part2 and select the Replace option. From the drop-down list select "LFILLET". Select "Yes" and Part2 will be deleted from the file. Your drawing should resemble Figure 7.7.

Figure 7.7

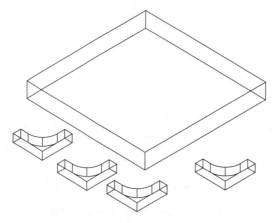

16. Save the file as `\Md2book\Drawings\Chapter7\Ex7-1.dwg`.

Bottom Up Approach

The bottom up assembly approach uses files that are referenced to an assembly file. If you have used xref's before, you will see a resemblance to the way that files are referenced to an assembly. Even though the procedure resembles xref's, do not use the XREF command. To create a bottom up assembly, create the parts in their individual files. If there are multiple parts in single file, they will be brought in as a subassembly. Any drawing views that were created in these files will not be brought into the assembly file. After the parts are created, start a new drawing, which will become the assembly file. In the new drawing, issue the Part Catalog command (AMCATALOG) or right-click in the white area of the browser and select Attach Definition. The Assembly Catalog dialog box will appear. Select the External tab if it is not already visible (see Figure 7.2). This tab is divided into two main sections. The Part and Subassembly Definitions section is on the left side, which shows the files in the current directory. The right side, Directories section, shows the directories where the files will be found. Not all the directories where the external parts are located will necessarily be shown by default. You can add individual files from directories not listed by using the Browse and Attach option, which will be discussed in the next section. To add or remove directories from the list, right-click on the All Devices folder. When this is done, you will get the following choices, as seen in Figure 7.8.

Figure 7.8

Add Directory: A browser will appear, allowing you to select the directory to add. You can repeat the procedure to add multiple directories, but only one directory can be current at a time.

Release Directory: Right-click on a directory name in the dialog box. It will be removed from the current set of directories.

Browse & Attach: Allows you to select individual files from any directory.

Release All: Removes all directories from the existing set of directories.

Include Subdirectories: Includes all subdirectories in the directory listing. This option needs to be set before you add a directory.

After adding a directory, or directories, make one of the directories current by selecting it with the left mouse button. The files in that directory will appear on the left side of the dialog box.

Viewing

To view all the files in all the added directories, left-click on All Devices from the top of the directory listings. To see an image of a file, left-click on the file name and a thumbnail image will appear in the preview window. This preview of the image shows the file as it appeared on the screen when it was last saved.

Sorting

To sort the files alphabetically or chronologically, right-click in the white area of the Part and Subassembly Definitions section and select alphabetical or chronological.

Inserting

To insert the file, you can either double-click on the file name or right-click and select Attach. You will then be returned to the drawing area to select a point where you want the part inserted. To insert another instance of the part, select another point, or press to return to the Assembly Catalog dialog box. The gray background will be removed from the file that was inserted.

Editing

To edit an external assembly or part, there are two methods. The first method is to open the part file in Mechanical Desktop, edit the part, update the part and save the part. Then open the assembly file; the changes to the part will be reflected in the assembly. The second method is to localize the part using the Part Catalog command (AMCATALOG). A localized part will have no relationship to the external part file. To localize a part, select the part name in the Assembly Catalog dialog box with the right mouse button and then select Localize.

Externalizing

There are times when you would like to take a local part and write it to its own file for detailing etc. This is referred to as externalizing. An externalized part is automatically referenced to the assembly file. To externalize a part, issue the Part Catalog command (AMCATALOG) and select the All tab. On the right side of the dialog box there will be a list of all local parts. Select the part to externalize with the right mouse button and then choose Externalize from the pop-up menu. Select a directory and name for the file. If there are drawing views associated with the part, they will not go with the externalized part; only the part information or geometry is externalized.

> **Notes:** The important things to remember are that a file can be attached to an unlimited number of files and that only the part information is inserted in the assembly. No drawing view information will come across.
>
> A single file can be attached to an unlimited number of assemblies.
>
> Changes made to an external part will be reflected in the assembly when it is next opened or updated.
>
> An assembly file can have both local and external parts.

Assemblies

Tutorial 7.2—Bottom Up Assembly

1. Start a new file.
2. Switch to an isometric view.
3. Issue the Part Catalog command (AMCATALOG).
4. Select the External tab.
5. From the Directories area, add the directory where the Chapter 7 files are located.
6. From the files in the Part and Subassembly Definitions section, insert one Base, one Gasket and one Cover in the file. Either double-click on the file name or right-click and select Attach.
7. Insert six bolts from the file HHCS1 file: again double-click on the file HHCS1 or right-click on the file and select Attach.
8. Save the file as \Md2book\Drawings\Chapter7\Ex7-2.dwg.
9. Open the file \Md2book\Drawings\Chapter7\Cover.dwg and create a 1" diameter hole through the center of the part. Update and save the part.
10. Open the file \Md2book\Drawings\Chapter7\Ex7-2.dwg. There should now be a hole through the center of the cover.
11. Through the Part Catalog command, localize the cover.
12. Make the cover part active by expanding Cover_1 and double-clicking on Part1_1. Then erase the hole in the center of the part. Update the part.
13. Through the Part Catalog command, externalize the cover and save it as \Md2book\Drawings\Chapter7\Cover2.dwg.
14. Save the file.
15. Open both files, \Md2book\Drawings\Chapter7\Cover.dwg and \Md2book\Drawings\Chapter7\Cover2.dwg, to verify that they are different.

Assembly Constraints

In the previous sections you learned how to insert multiple parts in an assembly file, but the parts did not have any relationship to each other. For example, if a bolt was placed in a hole, and the hole moved, the bolt would not move to the new hole position. Assembly constraints are used to create relationships between parts. So, if the hole moves, the bolt will move to the new hole location. In Chapter 1, you learned about geometric constraints. When geometric constraints are applied, they remove the number of dimensions, or constraints, required to fully constrain a profile. When assembly constraints are applied, they reduce the degrees of freedom (or DOF) that the parts can move freely in space. There are six degrees of freedom; three translational and three rotational. Translational means that a part can move along an axis: X, Y or Z. Rotational means that a part can rotate about an axis: X, Y or Z. As assembly constraints are applied, the degrees of freedom will be decreased. The first part created or added to the assembly will have zero degrees of freedom. This is the base component, also referred to as grounded. Other parts will move in relation to this part. To see a graphical display of the degrees of freedom remaining on a part, select a part's name in the browser with the right mouse button, then choose the DOF symbol from the pop-up menu. An icon and number will appear in the center of the part that shows the degrees of freedom remaining on a part. The number represents the order in which the part appears in the browser. The color of the DOF symbol is controlled by the color of grips and can be altered with the DDGRIPS command. To see the degrees of freedom of a part in a text format, issue the List Part Data command (AMLISTASSM) from the Assembly Modeling toolbar, as shown in Figure 7.9. After issuing the command, select a part and you will get a listing of the degrees of freedom remaining for the part.

Figure 7.9

Assemblies

The assembly modeler uses a variational solver, which allows parts to be assembled with no regard to the order of the parts. The first part in the browser is the base part. All parts are constrained to the base part. To reorder a part in the browser, select its name in the browser with the left mouse button and, keeping the button depressed, move the part up or down until it is in the new location. Like profiles, assemblies do not need to be fully constrained.

When constraining parts to one another, you will need to understand the terminology used. A list of terminology that is used with assembly constraints follows.

Line: Can be the center of an arc or circle, a selected edge or work axis.

Normal: A vector that is perpendicular to the outside of a plane. With Mechanical Desktop you can flip the normal direction once the plane is selected.

Plane: Can be defined by the selection of a plane or face, two non-collinear lines, three non-linear points, one line and a point that does fall on the line. When edges and points are used to select a plane, this is referred to as a construction plane.

Point: Can be an endpoint or a midpoint of a line or the center of an arc or circle.

Offset: The distance between two selected lines, planes, points or any combination of the three.

Types of Constraints

Mechanical Desktop has four types of assembly constraints: Mate, Flush, Insert and Angle. They can be accessed through the Constraints command (AMCONSTRAIN) on the Assembly Modeling toolbar or selected from their individual icons on the same flyout. If the Constraints icon is selected, the toolbar will float on the graphic screen until closed. Figure 7.10 shows the location of the Constraints command; to the left of the icon are

Figure 7.10

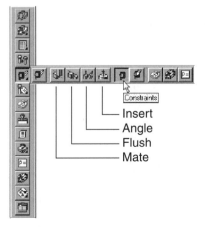

the individual constraints. Figure 7.11 shows the dialog box that will appear after the Constraints command is issued. This dialog box can be used to apply and edit assembly constraints. You can also apply the constraints from each individual icon and no dialog box will appear.

Figure 7.11

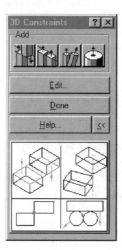

Mate: There are three types of mate constraints: plane, line and point.

Mate plane: assembles two parts so that the normals on the selected planes will be opposite one another when assembled.

Mate line: assembles to line or center lines of arcs and circular edges together.

Mate point: assembles two points (center of arcs and circular edges, endpoints and midpoints) together.

Flush: The normals on the selected planes will be pointing in the same direction when assembled.

Insert: Select the circular edges of two different parts and the centerlines of the parts will be aligned and a mate constraint will also be applied to the planes defined by the circular edges. The Insert constraint takes away five degrees of freedom with one constraint. It only works with parts that have circular edges. Circular edges define a centerline/axis and a plane.

Angle: You will specify the degrees between the selected planes.

The figures below show geometry before and after assembly constraints are applied.

Assemblies

Figure 7.12
Before applying mate constraint

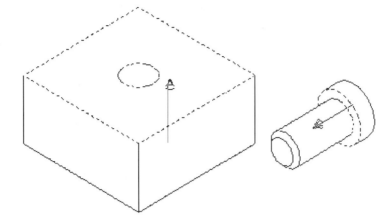

Figure 7.13
After applying mate constraint

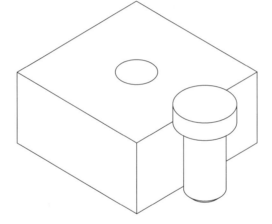

Figure 7.14
Before applying flush constraint

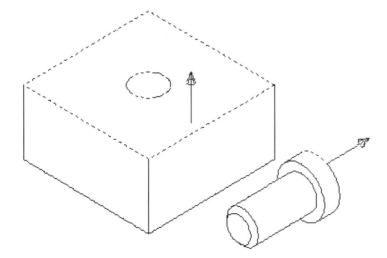

Figure 7.15
After applying flush constraint

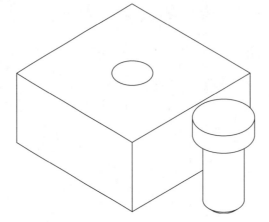

Figure 7.16
Before applying insert constraint

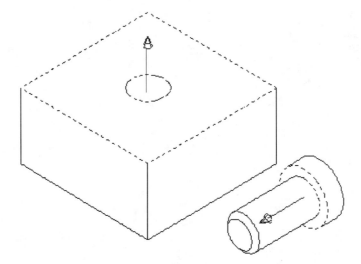

Figure 7.17
After applying insert constraint

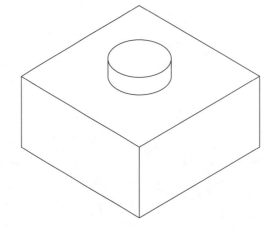

Assemblies

Figure 7.18
Before applying angle constraint

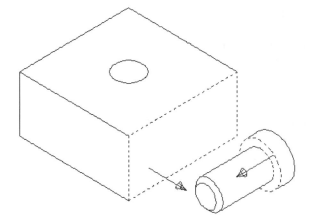

Figure 7.19
After applying angle constraint

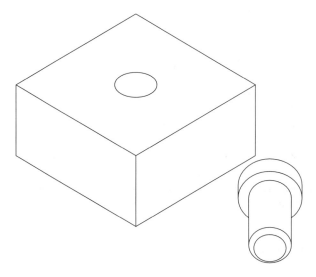

After selecting the assembly constraint type you want to apply, you will be prompted to:

```
Select first set of geometry:
```

Depending on the geometry you select, Mechanical Desktop will solve for that selection and allow you to cycle through all the possibilities or to select another edge or point to define the plane. If you select in the middle of a plane, you can cycle through the planes. If you select on an edge, you will cycle through the endpoints, midpoint, the line or edge itself and the two planes that are adjacent to the edge.

After you make the first selection, an icon representing a mouse will appear on the screen. On the screen the left mouse button will be flashing red. If you left-click, the geometry will be highlighted on the screen, showing you the possible choices. If the pick location could be on a separate part, it too will be cycled through. This object cycling is referred to as Intelligent Constraints. When the correct object is highlighted, press [Enter] or click the right mouse button.

You can also define a construction plane by selecting an edge or point and then moving the mouse over another line or point and selecting it. Continue with this technique until a plane is shown and then press [Enter].

To select a plane that is tangent to an arc or circle, select an isometric line that is closest to the quadrant you want. The quadrant is referenced to the world coordinate system. Cycle through with the left mouse button until the correct object is highlighted, then press [Enter], or select another object that will define a plane. The next prompt will ask you to:

```
Select second set of geometry:
```

Follow the same steps as for the first set. After defining the second set, you will be prompted for an offset distance. Type in a number and press [Enter]. If the geometry does not automatically update, select the Update Assembly icon on the bottom right side of the browser. Under the Assembly preferences, you can set Mechanical Desktop to automatically Update Assembly as Constrained, which is the default. If this is not checked, it will leave the parts in place until the assembly is updated manually. If the assembled parts make it hard to select new points, use the AutoCAD move command to reposition the parts and then update the assembly. The parts will be moved back to their constrained positions.

 Notes: You can create theoretical or construction planes or edges by moving the mouse cursor over an edge and cycling through the geometry and then moving it to the next edge and cycling through in the same manner until the plane or edge is defined. Then press [Enter].

A work plane can be used as a plane with assembly constraints.

A work axis can be used to define a line.

Assemblies

Tutorial 7.3—Constraining with Mate

1. Open the file \Md2book\Drawings\Chapter7\Ex7-3.dwg.

2. Turn on the degrees of freedom for both side plates by selecting their names in the browser with the right mouse button and selecting the DOF symbol. Watch the symbol change as constraints are applied.

3. Issue the Constraints command (AMCONSTRAIN) with the Mate option or select the Mate Constraint command (AMMATE).

4. Select the bottom plane of the side plate and the back plane of the base, as shown in Figure 7.20. You may need to cycle through with the left mouse button. For clarity, the figure is shown with the DOF symbol turned off.

Figure 7.20

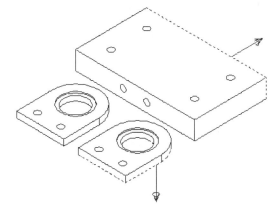

5. Press [↵Enter] to accept the default offset distance of "0.00".

6. Press [↵Enter] to apply another Mate constraint.

7. Select the back left hole of the base plate and the upper right hole of the side plate, as shown in Figure 7.21. Your side plate may be in a different position on the screen from that shown. For clarity, the figure is shown with the DOF symbol turned off.

8. Press [↵Enter] to accept the default offset distance of "0.00".

9. Press [↵Enter] to apply another Mate constraint.

10. Select the back right hole of the base plate and the left hole of the side plate, as shown in Figure 7.22. Your side plate may be in a different position on the screen from that shown.

Figure 7.21

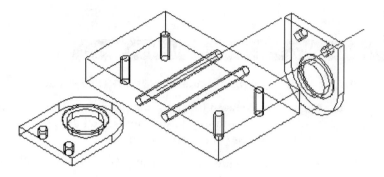

Figure 7.22

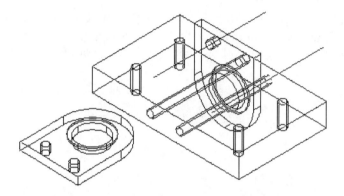

11. Repeat the above steps to assemble the other side plate to the Base. When you are done, your drawing should look like Figure 7.23.

Figure 7.23

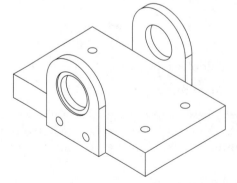

12. Save the file.

Assemblies

Tutorial 7.4—Constraining with Flush

1. Open the file \Md2book\Drawings\Chapter7\Ex7-4.dwg.

2. Issue the Constraints command (AMCONSTRAIN) with the Mate option or select the Mate Constraint command (AMMATE).

3. Select the bottom plane of the left guide and the top plane of the base, as shown in Figure 7.24. You may need to cycle through with the left mouse button.

Figure 7.24

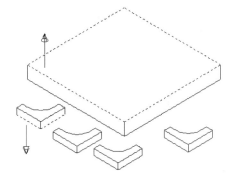

4. Press [Enter] to accept the default offset distance of "0.00".

5. Issue the Constraints command with the Flush option or select the Flush Constraint command (AMFLUSH).

6. Select the front left face of the guide and the left back of the base, as shown in Figure 7.25.

7. Press [Enter] to apply another Flush constraint.

Figure 7.25

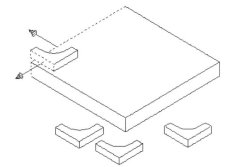

231

8. Select the front left face of the guide and the left front of the base, as shown in Figure 7.26.

Figure 7.26

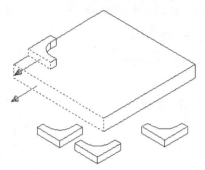

9. Press [⏎] to accept the default offset distance of "0.00".
10. Constrain the other three guides with the Mate and Flush constraints. When you are done, your drawing should look like Figure 7.27.

Figure 7.27

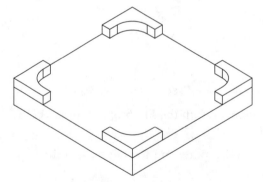

11. Save the file.

Assemblies

Tutorial 7.5—Constraining with Insert

1. Open the file \Md2book\Drawings\Chapter7\Ex7-5.dwg.

2. Issue the Constraints command (AMCONSTRAIN) with the Insert option or select the Insert Constraint command (AMINSERT).

3. Select one of the top circles of the base and a circle on the bottom of the gasket, as shown in Figure 7.28. Since the circles are concentric, it does not matter which circle in the plane is selected.

Figure 7.28

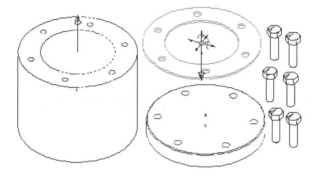

4. Press [Enter] to accept the default offset distance of "0.00".

5. Issue the Constraints command with the Mate option or select the Mate Constraint command (AMMATE).

6. To align the holes of the base and gasket; select one of the circles from the top of a hole in the base. The centerline should be highlighted. If it is not, cycle through until it is and press [Enter].

7. Select a hole from the outside perimeter of the gasket and cycle through until the centerline is highlighted and press [Enter].

8. Press [Enter] to accept the default offset distance of "0.00".

9. Follow steps the same steps to insert and align the cover.

10. Then use the Insert constraint to align the six bolts into the six holes of the cover. When you are done, your drawing should look like Figure 7.29. Since we do not care how the bolts are rotated in the holes, we will leave them under-constrained.

Figure 7.29

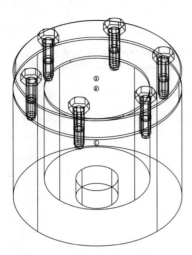

 Tutorial 7.6—Constraining with Insert and Angle

1. Open the file \Md2book\Drawings\Chapter7\Ex7-6.dwg.

2. Issue the Constraints command (AMCONSTRAIN) with the Insert option or select the Insert Constraint command (AMINSERT).

3. Select the top left arc of the bottom arm and the lower left arc of the left arm, as shown in Figure 7.30.

Figure 7.30

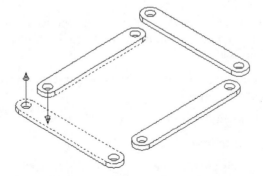

4. Press [⏎] to accept the default offset distance of "0.00".

5. Press [⏎] to repeat the Insert constraint.

6. Select the top right arc of the bottom arm and the lower left arc of the right arm, as shown in Figure 7.31.

Figure 7.31

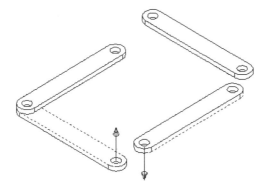

7. Press [⏎] to accept the default offset distance of "0.00".
8. Press [⏎] to repeat the Insert constraint.
9. Select the bottom right arc of the right arm and the top right arc of the top arm, as shown in Figure 7.32.

Figure 7.32

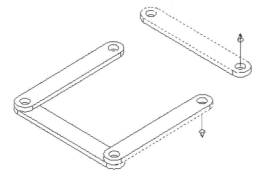

10. Press [⏎] to accept the default offset distance of "0.00".
11. Turn on the DOF symbol for Part1_4.
12. Use the Mate constraint with the line option to align the center of the left arm and the center of the top arm, as shown in Figure 7.33.
13. Use the Angle Constraint command (AMANGLE) to fully constrain the assembly. Select the inside plane of the bottom arm and the inside plane of the left arm, as shown in Figure 7.34.
14. Type in a value of "-30" and your drawing should look like Figure 7.35.
15. Save the file.

Figure 7.33

Figure 7.34

Figure 7.35

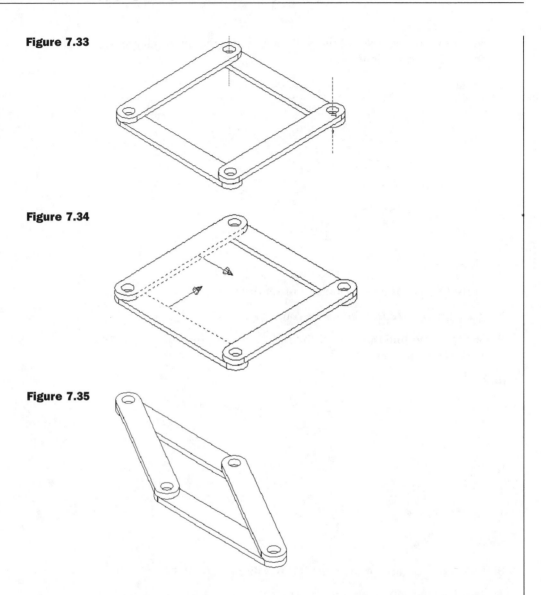

Assemblies

Tutorial 7.7—Constraining with Point and Line

1. Open the file \Md2book\Drawings\Chapter7\Ex7-7.dwg.
2. Issue the Constraints command (AMCONSTRAIN) with the Mate option or select the Mate Constraint command (AMMATE).
3. Select the top plane of the base (green) and the bottom plane of Part2 (magenta), as shown in Figure 7.36.

Figure 7.36

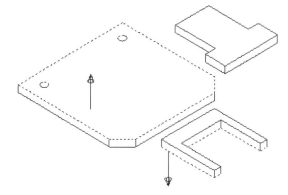

4. Press [⏎] to accept the default offset distance of "0.00".
5. Press [⏎] to repeat the Mate constraint.
6. Select the top back edge of Part1 (green) and cycle through until the middle point is highlighted. Do the same for the back bottom edge of Part2 (magenta), as shown in Figure 7.37.

Figure 7.37

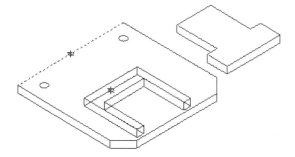

7. Press [↵] to accept the default offset distance of "0.00".
8. Press [↵] to repeat the Mate constraint.
9. Select the top plane of part1 (green) and the bottom plane of Part3 (red), as shown in Figure 7.38.

Figure 7.38

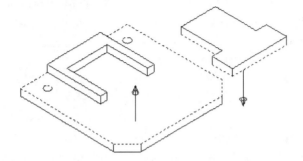

10. Press [↵] to accept the default offset distance of "0.00".
11. Press [↵] to repeat the Mate constraint.
12. Select the middle plane of Part2 (magenta) and the back plane of Part 3 (red), as shown in Figure 7.39.

Figure 7.39

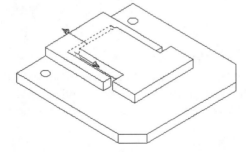

13. Type in an offset distance of ".5".
14. Press [↵] to repeat the Mate constraint.
15. Select the front edge of Part2 (magenta) and the front edge of Part 3 (red), as shown in Figure 7.40.
16. Press [↵] to accept the default offset distance of "0.00" and your drawing should look like Figure 7.41.
17. Save the file.

Assemblies

Figure 7.40

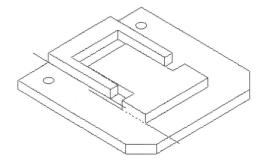

Figure 7.41

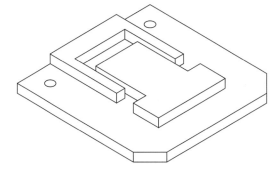

Tutorial 7.8—Constraining with Line/plane and Point

1. Open the file `\Md2book\Drawings\Chapter7\Ex7-8.dwg`.

2. Issue the Constraints command (AMCONSTRAIN) with the Mate option or select the Mate Constraint command (AMMATE).

3. Select the top plane of the base (green) and the bottom plane of Part2 (magenta), as shown in Figure 7.42.

Figure 7.42

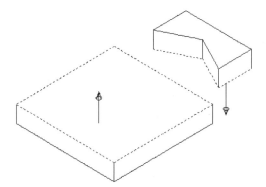

4. Press [Enter] to accept the default offset distance of "0.00".
5. Press [Enter] to repeat the Mate constraint.
6. Create a construction plane by selecting the leftmost vertical edge, cycling through until the line is highlighted; do not press [Enter]. Then move the mouse to the opposite right vertical edge and select it. Press [Enter] to accept this middle construction plane. When complete, your drawing should look like Figure 7.43.

Figure 7.43

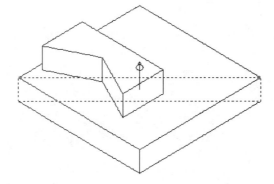

7. Select the two front vertical edges in Part2 to define its plane as well. Your drawing should look like Figure 7.44.

Figure 7.44

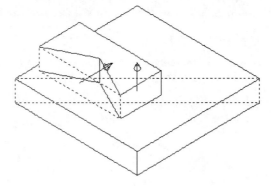

8. If the normal arrow is not pointing inward, flip it and press [Enter].
9. Press [Enter] to accept the default offset distance of "0.00".
10. Change to a plan view (**9**). Your drawing should look like Figure 7.45.
11. Change to an isometric view (**88**) and press [Enter].

Figure 7.45

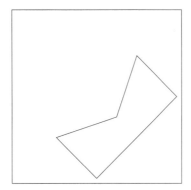

12. Issue the Constraints command (AMCONSTRAIN) with the Mate option or select the Mate Constraint command (AMMATE).

13. Select the top front edge of Part1 (green), cycle through until the point is highlighted and then press [↵]. Select the bottom front edge of Part2 (magenta), cycle through until the point is highlighted and then press [↵]. Figure 7.46 shows the points.

Figure 7.46

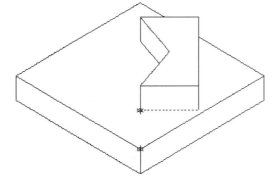

14. Type in an offset distance of ".5" and press [↵]. Your drawing should look like Figure 7.47.

15. Save the file.

Figure 7.47

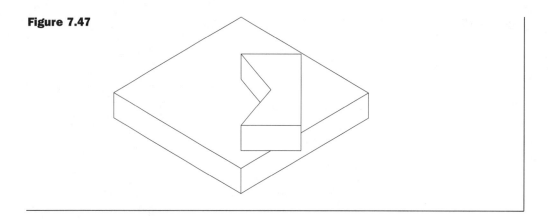

Editing Assembly Constraints

After an assembly constraint has been placed, you may want to edit or delete it to correctly position the parts. There are three methods for editing or deleting assembly constraints.

The first two methods are through the browser. In the browser, expand the part name and you will see the assembly constraints. Double-click on the constraint name and an Edit 3D Constraint dialog box will appear, allowing you to edit the constraints. The second method is to right-click on the constraint and a pop-up menu will appear. Choose either Edit or Delete. If you choose Delete, the constraint will be deleted from the part. If you choose Edit, an Edit 3D Constraint dialog box will appear. In the dialog box, type in a new value and update the assembly by selecting the Update icon (lightning bolt) in the same dialog box.

The third method is to select the Edit Constraints command (AMEDITCONST) from the Assembly Modeling toolbar. After you issue the command, the Edit 3D Constraints dialog box will appear. On the left side of the dialog box you will see a listing of the parts that have constraints. When a part name is selected, it will highlight in the drawing window. If you are unsure of a part's name, select the Select button and you will be returned to the drawing area, where you can select a part. You will then be returned to the dialog box. Once a part is highlighted, you can cycle through the constraints by selecting Show; two double arrows will appear. Cycle through the constraints by selecting on the arrows to move forward or backward. Below the arrows you will see the type of constraint, and the constraint will be visible in the drawing area. To edit the offset value, type in a new value from the Edit Constraints area. If the assembly does not automatically update, select the lighting bolt icon to re-assemble the parts. To delete the current constraint, select Current from the Delete Constraints area. To delete all the constraints from the current part, select All from the Delete Constraints area.

Assemblies

If you want to change a constraint to a different type, you need to first delete the original constraint and then apply the new constraint. If you do not delete the constraint first and you try to apply a different constraint in its place, you will get the message at the command line: Solve failed with the constraint name that was being added. For example:

Solve failed. Mate pl/pl constraint not added.

Figure 7.48

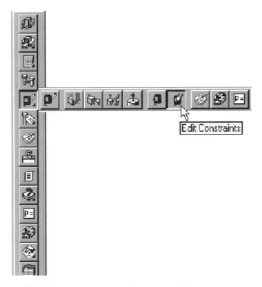

Figure 7.49

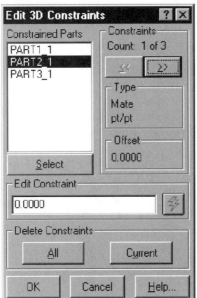

Tutorial 7.9—Editing Constraints

1. Open the file \Md2book\Drawings\Chapter7\Ex7-9.dwg.
2. Issue the Edit Constraints command (AMEDITCONST).
3. Select PART3_1.
4. Cycle to the third constraint and change the offset value from ".5" to "0". Update the part assembly if necessary. When complete, your drawing should look like Figure 7.50.

Figure 7.50

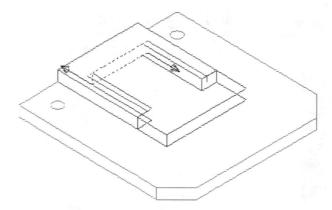

5. Delete the Mate ln/ln constraint for PART3_1.
6. Select OK to exit the command.
7. Apply a Mate constraint to the left inside plane of Part2 (magenta) and to the left outside plane of Part3 (red), as shown in Figure 7.51.

Figure 7.51

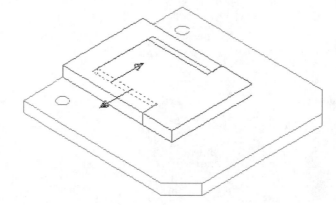

8. Press to accept the default offset distance of "0.00". When complete, your drawing should look like Figure 7.52.

Figure 7.52

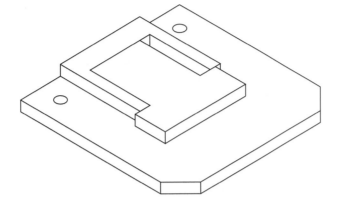

9. Save the file.

Interference Checking

To check the interference between parts, issue the Check Interference command (AMINTERFERE) from the Assembly Modeling toolbar as shown in Figure 7.53. This command checks interference between two sets of geometry. After issuing the command you will be prompted:

```
Nested part/subassembly selection? Yes/<No>:
```

If you select yes, you will select the subassembly that you want to work with. If you press ⏎, you will be prompted to select the first set. You can select a single part or multiple parts using any AutoCAD selection method. When you are done selecting the first set, press ⏎. Then select the second set to be checked against the first set. If no interference is found, you will see a message at the command line:

```
Parts/Subassemblies do not interfere.
```

If interference is found, the names of the interfering parts will appear at the command line. You may need to flip screens to see the message. At the command line, you will see a prompt:

```
Create interference solids? Yes/<No>:.
```

If you press **Y** and ▭, a 3DSOLID will be created representing the interference. The 3DSOLID will be red and placed on the current layer. Another prompt will appear at the command line:

```
Highlight pairs of interfering parts/subassemblies? Yes/<No>:
```

If you press **Y**, the interfering parts will be highlighted in the drawing area. If there are multiple interferences, press **N** to advance to the next set. To exit the command, press **X**.

After exiting the command, you can move the solid or interference and use the AutoCAD distance command along with object snaps to define the amount of interference. The distance reported back is accurate to the number of decimal places set through the DDUNITS command. Mass property information can also be checked. After finding out the amount of interference, go back and edit the part(s) to fix the interference. Another option is to use the Combine command (AMCOMBINE) to remove or add the 3DSOLID that was created after the interference was found. The down side to this method is that the solid or interference is a 3D solid and not a parametric part and it cannot easily be changed.

Figure 7.53

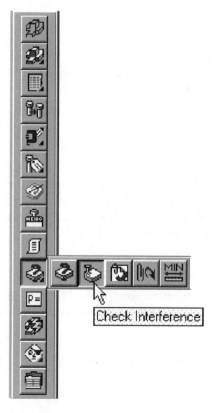

Assemblies

Tutorial 7.10—Interference Checking

1. Open the file \Md2book\Drawings\Chapter7\Ex7-10.dwg.
2. Issue the Check Interference command (AMINTERFERE).
3. For the prompt:

   Nested part/subassembly selection? Yes/<No>:

   press [Enter] to accept the default.
4. Select the outside cylinder for the first set and press [Enter].
5. Select the inside pin and press [Enter] for the second set.
6. Press the **F2** key on your keyboard to see that interference was found in both Part1 and Part2. Press the **F2** key to return to the drawing screen.
7. Press **Y** and [Enter] to create an interference solid.
8. Press **Y** and [Enter] to highlight the interfering pairs.
9. Press **X** and [Enter] to exit the command.
10. Use the AutoCAD move command with the "L" option to move the last object created (the interfering solid) to the side.
11. Use the AutoCAD distance command with the quadrant object snap to find how much interference exists.
12. Change one of the parts diameters to remove the interference.
13. Issue the Check Interference command (AMINTERFERE) and recheck for interference. If there is still interference, go back through steps 11 to 13 until there is no interference.
14. Save the file.

Creating Scenes

After creating the assembly file, you may want to show all of the parts in different positions, like an exploded view, or hide specific parts for a drawing view. You can create scenes to show parts in different positions and to create drawing views of an assembly. There is no limit to the number of scenes in a file. To create a scene, issue the Create Scene command (AMNEW) icon from the Scenes toolbar (as shown in Figure 7.54) or select the Scene Tab from the browser, right-click in a blank area in the browser and choose New Scene from the pop-up menu.

Figure 7.54

After issuing the command, type in a name for the scene and press [←Enter]. Then type in a value for the explosion factor and press [←Enter]. The explosion factor will be the distance between the parts. Any number is valid for the explosion factor, including zero. Press [←Enter] again to make this the active scene. This automatic movement only works for parts that have been constrained with the Mate or Insert assembly constraints. The Mate and Insert constraints have the normals facing toward each other. Think of them with regard to scenes as two magnets that are put together. As the magnets come together, they want to push away from one another. The part that appears first in the browser will be the part that stays stationary; the other part will move away from it. If the parts do not automatically explode, there could be one of three reasons:

1. The parts were not constrained with assembly constraints.
2. There was no Mate or Insert constraint used.
3. There was more than one Mate or Insert constraint used for a single part. If multiple Mates or Inserts were used on a single part, there could be two directions that the part could move. The result is that the part stays stationary.

If a part does not automatically move, you can use the Add Tweaks command (AMTWEAK) to move the parts. Tweaks will be covered in the next section. To switch between multiple scenes in a file, issue the Create Scene command (AMNEW) and select the scene name from the drop-down list, or double-click on the scene name in the browser under the Scene tab. To tell which scene is active, look in the browser under the Scene tab (the current scene's name will be white and the inactive scenes gray), or look at the lower left corner of the AutoCAD screen, which will show the current scene name.

After you create a scene, the parts will be held in a block and cannot be edited while a scene is current. To edit a part, select on the Assembly tab in the browser. Parts will revert back to the their positions before the scene was created. In this assembly mode, parts and assembly constraints can be added, edited or deleted. After editing the parts, make a scene active and all the changes will be reflected in that scene.

Tweaking Parts in a Scene

Tweak is the term used to describe the repositioning of parts in a scene. Once a scene is created, issue the Add Tweaks command (AMTWEAK) from the Scenes toolbar, as shown in Figure 7.55, or right-click on the part name in the browser under the Scene Tab and choose Add Tweak from the pop-up menu. Then select the part to tweak and a dialog box will appear, offering three choices: Move, Rotate and Transform.

Move: Moves a part in the direction of an edge of a part that you select. After selecting Move, you will be prompted to:

```
Select reference geometry:
```

Select an edge. The part will then be moved in the direction and distance of the selected edge.

Rotate: Rotates a part around an edge of a part that you select. After selecting Rotate, you will be prompted to:

```
Select reference geometry:
```

Select an edge and then type in a value for the angle of rotation. The part will be rotated around the selected edge.

Transform: Moves or rotates a part about the XYZ axis of the current UCS or you can select two points to determine the direction. Type in a value and an arrow will appear to verify if you want to go in the positive or negative direction.

The base part in an assembly cannot be tweaked. Once a tweak has been created, it cannot be edited, it can only be deleted. To delete a tweak, issue the Delete Tweaks command (AMDELTWEAKS) from the Scenes toolbar, as shown in Figure 7.56, and select a part. If a part has multiple tweaks, they will all be deleted.

Figure 7.55

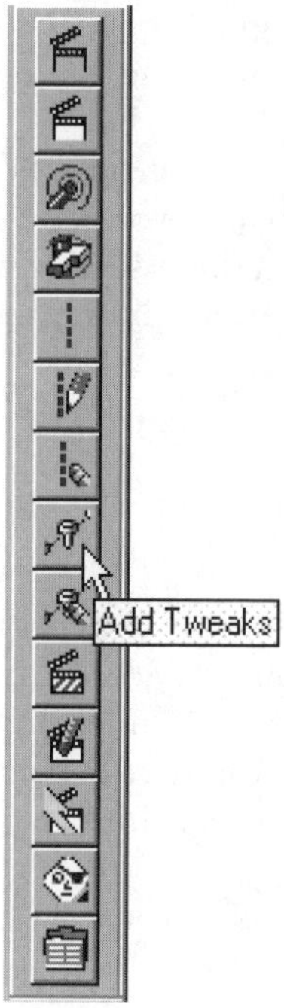

Figure 7.56

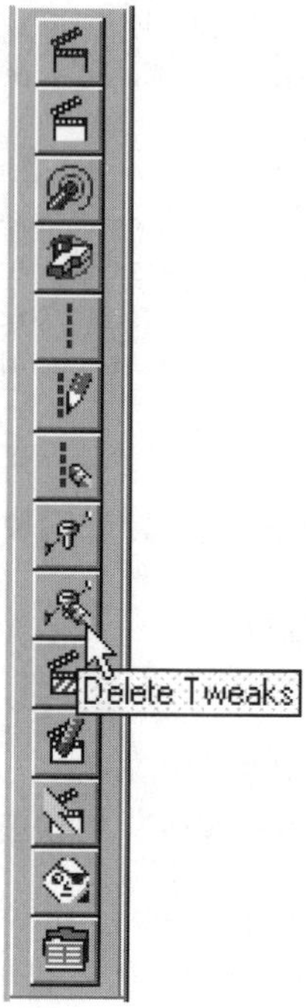

Creating Trails

To add a line that shows how a part has moved in a scene, issue the Create Trail command (AMTRAIL) from the Scenes toolbar, as shown in Figure 7.57, or right-click on the part name in the browser under the Scene Tab and choose Add Trail from the pop-up menu. You will be prompted to:

```
Select reference point on part/subassembly:
```

Select a part. Where you select the geometry is where the trail will start. When an arc or circular edge is selected, the trail will go to the center of the arc or circular edge. If a line is selected, the trail will start at the closest endpoint. After the geometry has been selected, a dialog box will appear, as shown in Figure 7.58. The dialog box contains two sections: Offset at Current and Assembled Position. The Current position refers to the part's place in the scene. The Assembled position refers to the part's position with a zero explosion factor. The Over Shoot and Under Shoot refer to the distance that a trail should go beyond or stop before the part's position.

Editing a Trail

When creating trails, it is easier to accept the zero defaults and then edit them if they do not look correct. The trails will take on the properties of the Am_tr layer. This would be a good item to change in your template file.

Assemblies

Figure 7.57

Figure 7.58

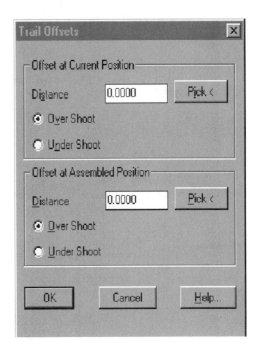

To edit a trail, issue the Edit Trail command (AMEDITTRAIL) from the Scenes toolbar, as shown in Figure 7.59, or right-click on the part name in the browser under the Scene Tab and select Edit Trail. You will be prompted to:

```
Select trail to edit:
```

Select the trail to edit and the same dialog box that was used to create the trail will reappear. Type in different values until the trail looks correct.

Figure 7.59

Mechanical Desktop 2.0: Applying Designer and Assembly Modules

Deleting a Trail

To delete a trail, issue the Delete Trail command (AMDELTRAIL) from the Scenes toolbar or right-click on the part name in the browser under the Scene Tab and select Delete Trail. You will be prompted to:

```
Select trail to delete:
```

Select the trail and it will be deleted.

Figure 7.60

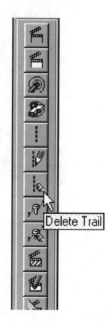

Tutorial 7.11—Creating Scenes, Tweaks and Trails

1. Open the file \Md2book\Drawings\Chapter7\Ex7-11.dwg.

2. Issue the Create Scene command (AMNEW) from the Scenes toolbar to create a new Scene.

3. Type in a name "Exploded" and press [⏎]. Type in a value of "2" for the explosion factor and press [⏎]. Press [⏎] to make this the active scene. When complete, the parts should look like Figure 7.61.

4. Press [⏎] to return to create another scene with a name "Assembled", with an explosion factor of "0", and make it the active scene. When complete, the parts will be in an assembled position.

Figure 7.61

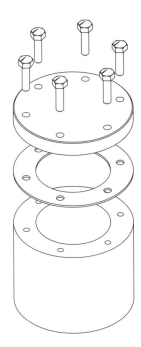

5. Make the Exploded scene current and change the explosion factor to "3".
6. Issue the Add Tweaks command (AMTWEAK), select the rightmost bolt and select Transform from the dialog box.
7. Press in **M** and [←Enter] to move the part.
8. Press in **X** and [←Enter] to move the part along the X axis.
9. Type in "2" for the distance, press [←Enter] and press [←Enter] again to accept the default direction.
10. Press [←Enter] to move the part again.
11. Press **Z** and [←Enter] to move the part along the Z axis.
12. Type in "2" for the distance, press [←Enter] and press [←Enter] again to accept the default direction.
13. Press **X** and press [←Enter] to exit the command.
14. Issue the Create Trail command (AMTRAIL) and select the bottom of the bolt that was tweaked. Accept the default values in the dialog box.
15. Issue the Edit Trail command (AMEDITTRAIL) and edit the trail so that the Under Shoot at the assembled position is "1".

16. Change the linetype of the Am_tr layer to center and change the color to red. When complete, your drawing should look like Figure 7.62.

Figure 7.62

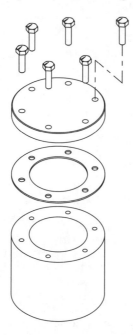

17. Practice tweaking the parts and adding trails.
18. When done practicing, save the file.

Exercises

For the following exercise, follow the instructions.

Exercise 1.1—Vice Assembly

1. Open the file \Md2book\Drawings\Chapter7\ViceAssy.dwg.
2. Assemble the parts as shown in Figure 7.63.
3. Create a scene with an explosion factor of "3".
4. Add tweaks and trails to all the parts, as shown in Figure 7.64.

Assemblies

 Note: When constraining the handle to the screw, align the center of the handle to the work axis of the screw. For the offset between the screw and the handle, use the Mate constraint and offset it to the bottom tangency of the circle of the screw, using a distance of "1.5".

Figure 7.63

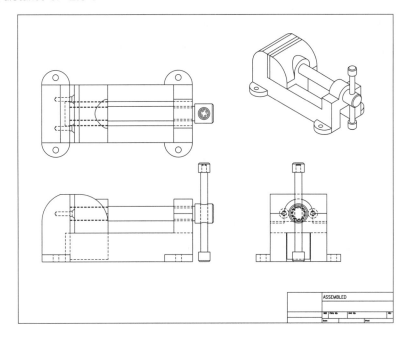

Figure 7.64

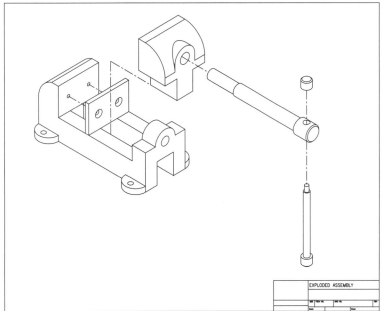

Review Questions

1. A bottom up assembly has all the parts in the same file. T or F?
2. When you create an assembly, all the files must either be entirely inside or entirely referenced to the assembly file. T or F?
3. Explain the difference between a Mate and a Flush constraint.
4. What do degrees of freedom refer to?
5. An assembly does not have to be fully constrained. T or F?
6. To change an assembly constraint to a different type (for example, Mate to Flush), the original constraint needs to be deleted first. T or F?
7. What does the Insert constraint do?
8. The solid that is created from the Check Interference command is parametric. T or F?
9. What are two purposes for creating scenes?
10. Parts that are constrained with the Mate or Insert constraint will automatically be exploded when a scene is created that has an explosion factor greater than zero. T or F?

chapter

Drawing Views and Annotations

After you create the part or assembly, the last step is to create drawing views. Even though you have a complete part, most companies still require traditional drawing views and in this chapter, you will learn how to create drawing views from individual parts and assemblies. Once the views are created, you will clean up the dimensions, add annotations and balloons and create a bill of materials to complete the drawing.

After completing this chapter, you will be able to:

- Create drawing views from a part as well as an assembly.
- Edit and move drawing views.
- Move, hide, break, join, align, insert, and modify text properties of dimensions.
- Add reference dimensions.
- Add annotations such as geometric, surface, weld symbols, datums and centerlines.
- Add balloons and create a bill of materials.

Creating Drawing Views

Before creating drawing views, you need to have a part or assembly. It does not need to be complete, since the part and drawing views are associative in both directions. This means that if the part changes, the drawing views will automatically be updated when you return to the drawing views. If a parametric dimension changes in a drawing view, the part will get updated before the drawing views get updated.

When you are working with an assembly, it is recommended that you create a scene before generating drawing views. To create a drawing view for the current part or assembly, issue the Create Drawing View command (AMDWGVIEW) from the Drawing Layout toolbar, as shown in Figure 8.1, or right-click in the browser under the Drawing tab and select Create View from the pop-up menu. Then a dialog box will appear, as shown in Figure 8.2. The dialog box is divided into several areas: type, data set, section views, hidden lines and scale.

Figure 8.1

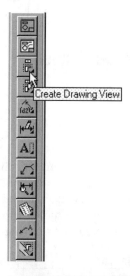

Figure 8.2

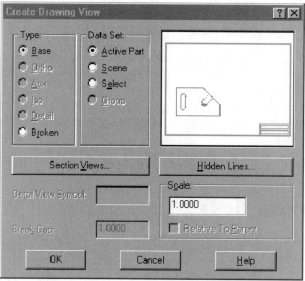

Type

Base: The first drawing view from a part or assembly, the base view is required before other views can be created. It is up to you to decide which view is the base view: top, front or side.

Ortho: A drawing view that is projected horizontally or vertically from another view.

Aux: A drawing view that is perpendicular to a selected edge of another view.

Iso: A 30° isometric view generated from any view. An isometric view can be projected to any of the four quadrants.

Detail: A selected area of an existing view will be generated at a specified scale.

Broken: A view in which the middle of the part is removed and the ends remain. The overall dimension will reflect the correct length of the part.

Depending on whether or not there are existing drawing views in a file, some options may be grayed out.

Data Set

Active Part: Only the active part will appear in the drawing view.

Scene: You will specify the scene in which the drawing view will be regenerated. This is used for creating views of an assembly.

Select: Only the selected parts and objects will appear in the drawing view. Once the entities have been selected, there is no way to add additional parts to the drawing view. It is recommended that you use groups if you want to select parts instead of creating scenes.

Group: After you create a group in AutoCAD, this will be a valid option. The parts in the selected groups will appear in the drawing view.

Section Views

After selecting this button, you will have a choice of creating the section views detailed below, along with the specified hatch pattern and symbol. A section view can be a base view or an ortho view. In Mechanical Desktop, you can also create a sectioned iso view.

Full: Creates a section view that will go straight through a specified point or selected work plane. When you select a point, the point will go to the center of a selected arc or circle or to the nearest endpoint of a selected line.

Half: Creates a section view that is perpendicular to the specified point. You will be prompted to flip the direction of the cut.

Offset: Creates a section view based on the selected cutting line.

None: No section view will be created.

Hatch Pattern: Check this area if you want the view hatched and to select the pattern.

Section Symbol: Specify the symbol to be used in the section view.

Hidden Lines

Calculate hidden lines: When checked, hidden lines will be calculated. Otherwise, all geometry will be a continuous linetype, appearing like a wireframe part.

Hide hidden lines: When checked, the hidden lines will disappear from the drawing views.

Linetype of hidden lines: Specify the linetype to use for hidden lines.

Display tangencies: When checked, tangency geometry will be displayed.

Scale

Specify the scale for the created view. When you create an Ortho or an Aux view, the scale will be grayed out because the scale is dependent on the view it is projected from.

When you activate the Drawing Layout toolbar or select the Drawing tab in the browser, you will automatically be placed in drawing mode or drawing manager. Drawing mode does switch to paperspace, but it is recommended that you switch to this mode with the Drawing tab because Mechanical Desktop manages the objects created here. All drawing views are created in paperspace. Insert your title block in paperspace at full scale, because you will plot at full scale (1=1). All layer creation and the choice of layer that objects are placed on are automatically handled by Mechanical Desktop. Figure 8.3 shows the layer names as well as what is placed on the layers. The color and the linetype of the layers can be set in a template file. Through the drawing preferences, you can set up color information and drafting standards and set third or first angle projection. The dimensions will take on the properties of the current dimension style. Set up the dimension styles in your template file to avoid having to do this work over again.

To create a base view, select Base for the Type and type in a scale. Select OK and you will be returned to the part or assembly. Then select a plane, face, work plane or UCS to align the view and orient the XY axis for the base view. This sequence will be similar to making a plane or work plane the active sketch plane. After the correct orientation is selected, press [⏎] and you will be returned to paperspace, where you can select a point to place the view. Keep selecting a point until the location is correct and then press [⏎]. The view will then be created at this location. If the scale or location is incorrect, you can edit or move the views later. To create a view based on another view, issue the Create Drawing View command (AMDWGVIEW) or select the view in the browser with the

Drawing Views and Annotations

Figure 8.3

Am_bl	=	Balloon information is placed on this layer.
Am_bm	=	Bill of materials information is placed on this layer.
Am_hid	=	Hidden lines are placed on this layer.
Am_pardim	=	Parametric dimensions are placed on this layer.
Am_ refdim	=	Reference dimensions are placed on this layer.
Am_tr	=	Trails are placed on this layer.
Am_views	=	View borders are placed on this layer.
Am_vis	=	Visible lines are placed on this layer.
Am_work	=	All work planes, work axis and work points are placed on this layer. Change the color of this layer to better see these workplanes when they are created.

right mouse button and choose Create View from the pop-up menu. A Create Drawing View dialog box will appear. Select the type of view (Ortho, Iso etc.) and select OK. You will be prompted to select a parent view if you issued the command from the icon. If the command was issue from the browser, the parent view is already identified. Select a point in the parent view from which to create this new view. This parent view does not have to be the base view. After selecting the parent view, you will be prompted for a location for the view. Select a point to place the new view. After the location is correct, press and the view will be created.

Note: For drawing views created from an assembly, no parametric dimensions will appear. You can add reference dimensions as required.

When creating an isometric view of an exploded assembly, you can create the isometric view from another view or from the orientation of the current view.

Mechanical Desktop 2.0: Applying Designer and Assembly Modules

Tutorial 8.1—Creating Drawing Views from a Part

1. Open the file \Md2book\Drawings\Drawings\Chapter8\Ex8-1.dwg and go to the Drawing tab.

2. To create a base view, issue the Create Drawing View command (AMDWGVIEW) or right-click in the browser and select Create View. In the Create Drawing View dialog box, change the scale to ".5" and select OK.

3. Select an edge that defines the plane as shown in Figure 8.4. Select one of the horizontal or vertical edges that define the plane to orient the XY so that the UCS looks like Figure 8.4 and then press ⏎.

Figure 8.4

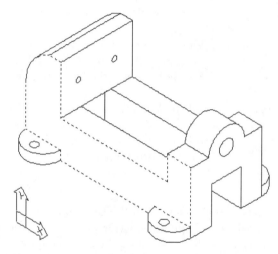

4. You will be returned to paperspace; select a point in the lower left corner of the existing border, as shown in Figure 8.5.

5. Press ⏎ to repeat the Create Drawing View command or right-click on the base view in the browser. Select Ortho, select OK, select a point in the front view and then select a point to the right so that the side view appears as shown in Figure 8.6.

6. Press ⏎ to repeat the Create Drawing View command. Select OK, select a point in the front view and then select a point to the top so that the top view appears as shown in Figure 8.7.

7. Press ⏎ to repeat the Create Drawing View command. Select Ortho, select Section Views and from the Section View dialog box select Full. Press **A** for the Section Symbol, select OK and then select OK in the first dialog box. Select a point in the top view, select a point to the right so that the section view will be

Drawing Views and Annotations

Figure 8.5

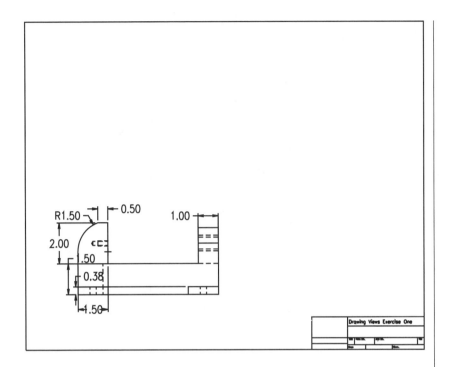

Figure 8.6

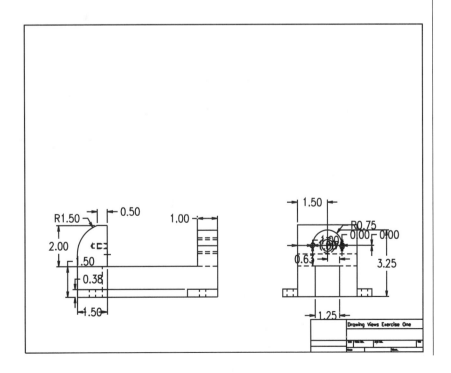

Figure 8.7

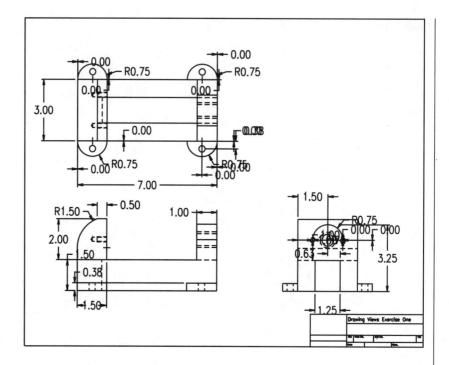

located as shown in Figure 8.8 and press [Enter] to accept this location. For the prompt:

```
Section through Point/Ucs/<Work plane>:
```

press **P** and press [Enter]. Select a point on the vertical line of the inside jaw (this is the line where the section line is drawn). Do not select on an arc. When complete, your drawing should look like Figure 8.8.

8. Press [Enter] to repeat the Create Drawing View command. Select Iso, change the scale to ".375", check relative to parent and then select OK.

9. Select a point in the front view and then locate the isometric view in the upper left corner of the drawing. When complete, your drawing should look like Figure 8.9.

10. Save the file.

Figure 8.8

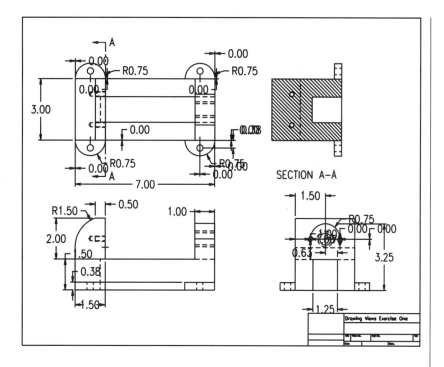

Figure 8.9

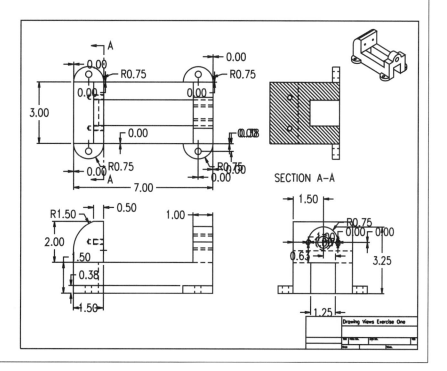

Mechanical Desktop 2.0: Applying Designer and Assembly Modules

Tutorial 8.2—Creating Drawing Views from an Assembly

In this file there are two scenes (assembled and exploded) that have already been created. The exploded scene has been tweaked and trails have been added.

1. Open the file \Md2book\Drawings\Drawings\Chapter8\Ex8-2.dwg.
2. Issue the Create Drawing View command (AMDWGVIEW). Select Base as the type, Scene for the Data Set, change the scale to ".5" and then select OK.
3. From the dialog box, select ASSEMBLED for the scene and select OK.
4. Select the top circular edge of the top cover for the plane and press [←Enter]. Orient the UCS by pressing **X** in reference to the world X.
5. Select a point in the upper left corner of the drawing and press [←Enter] to accept the position.
6. Create the front orthogonal view and an isometric view from this front view; create the isometric view ".75" scale relative to the parent. When complete, your drawing should look like Figure 8.10.

Figure 8.10

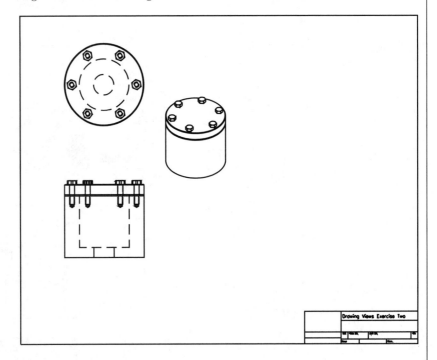

7. Make the EXPLODED scene current.

266

8. The viewpoint should already be in an isometric view. If it is not, change to an isometric view (**8**) and press [←Enter].

9. Change back to the drawing view by selecting the Drawing Layout icon or Drawing tab in the browser.

10. Issue the Create Drawing View command (AMDWGVIEW). Select Base as the type, Scene for the data set, change the scale to ".375", select Hidden Lines, check Hide Hidden Lines and then select OK.

11. From the dialog box, select EXPLODED for the scene and select OK or double-click on EXPLODED.

12. To use the current view as the plane, press **V** and [←Enter].

13. Press [←Enter] to accept the default orientation of the XY.

14. Locate the exploded view as shown in Figure 8.11.

Figure 8.11

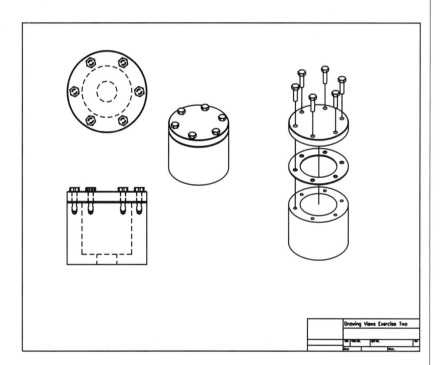

15. Save the file.

Editing Drawing Views

After creating the drawing views, you may need to edit the properties of a view. To edit a view's properties, issue the Edit Drawing command (AMEDITVIEW) from the Drawing Layout toolbar, as shown in Figure 8.12, or double-click on the view name in the browser under the Drawing tab and select Edit from the pop-up menu. After issuing the command, select a point in the view to edit and a dialog box will appear, as shown in Figure 8.13. Depending on the view that is selected, certain options may be grayed out. For example, if an orthogonal view is selected, the scale will be grayed out because it is dependent on the view from which it was projected.

Figure 8.12

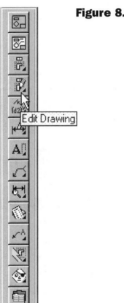

Figure 8.13

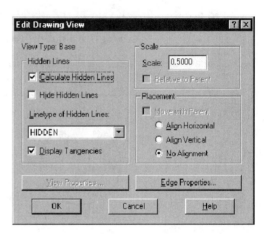

The properties discussed below can be changed through the dialog box.

Hidden Lines

Calculate Hidden Lines: When checked, Mechanical Desktop will calculate the hidden lines. This can be disabled as you create or edit a view and it can also be set through the Drawing preferences. For large or complex files, you can turn off the hidden line calculation when laying out the drawing views and turn it back on before plotting the file.

Hide Hidden Lines: When selected, the hidden lines will be removed from the view.

Linetype of Hidden Lines: Specify the linetype for hidden lines in this drawing view.

Drawing Views and Annotations

Display Tangencies: When checked, all tangencies will be displayed in the view.

View Properties: Only available when a detail view is selected; it will allow the detail boundary to be redefined and given a different symbol.

Scale

Scale: Available on base, detail and isometric views; all other views are tied to a parent view.

Relative to Parent: Available on isometric views. When checked, the view will be the percentage of the view from which it was created.

Placement

Move with Parent: Available on orthogonal views. When checked, the view will be aligned its parent view. If unchecked, the view can move freely. This option does not apply to base views.

Align Horizontal: When selected, you can realign a view along the horizontal axis that was moved away from its parent view. Use object snaps to realign the two views.

Align Vertical: When selected, you can realign a view along the vertical axis that was moved away from its parent view. Use object snaps to realign the two views.

No Alignment: When Move with Parent is unchecked, this option will be checked. This shows that the view can move freely.

Edge Properties: When selected, you will be prompted to either Unhide all edges that have been hidden with this command or press [←Enter] to select an edge(s) to modify. If you press [←Enter], select an edge or multiple edges using any AutoCAD selection technique. After selecting the edges, press [←Enter]. A dialog box allows you to hide the set or change the color, layer, linetype or linetype scale. When complete, select OK and you will be returned to the original dialog box. Then select OK in the original dialog box and the changes will take effect in the view.

Moving Drawing Views

To move a drawing view, issue the Move Drawing View command (AMMOVEVIEW) from the Drawing Layout toolbar, as shown in Figure 8.14, or right-click on the view name in the browser and select Move from the pop-up menu. If the command is issued from the toolbar, select a point in the view to move and left-click to select a point to move the view to. Keep selecting points until you are happy with the view placement and then press [←Enter], and the view will be moved to this new location. If you select a parent view, the children

Figure 8.14

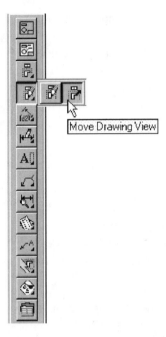

or dependent views will also move with it. Orthogonal views can only be moved along the axis in which they were created, unless you uncheck Move with Parent in the Edit Drawing View dialog box. Detail and isometric views can be moved freely.

Deleting Drawing Views

To delete a drawing view, issue the Delete Drawing View command (AMDELVIEW) from the Drawing layout toolbar, as shown in Figure 8.15, or right-click on the view name in the browser and select Delete from the pop-up menu. If the command is issued from the toolbar, left-click to select a point in the view to delete. If the selected view has a view that is dependent on it, you will be prompted:

```
View has # dependent views. Delete them also? Yes/No/<Cancel>:
```

If you press **Y** and [↲], the dependent view will be deleted. If you press **N** and [↲], only the selected view will be deleted. If you press [↲], the command will be cancelled and no view will be deleted. If the selected view has no dependent views, it will be erased without further action.

Figure 8.15

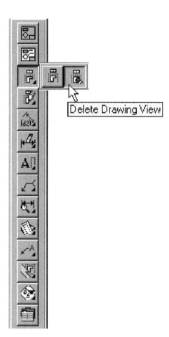

Tutorial 8.3—Editing, Moving and Deleting Drawing Views

1. Open the file \Md2book\Drawings\Drawings\Chapter8\Ex8-3.dwg or keep working on the file \Md2book\Drawings\Drawings\Chapter8\Ex8-2.dwg.

2. Issue the Edit Drawing command (AMEDITVIEW) or, in the browser, double-click on the name Base under the Assembled view.

3. Select a point in the top view of the assembled parts. Change the scale to ".5" and then select OK.

4. Press [⏎] to repeat the Edit Drawing command. Select a point in the isometric view of the assembled view, change the scale to ".75" and then select OK.

5. Press [⏎] to repeat the Edit Drawing command. Select a point in the exploded isometric view, check the box Hide Hidden Lines and then select OK.

6. Press [⏎] to repeat the Edit Drawing command. Select a point in the exploded isometric view, select Edge Properties and press [⏎] to select geometry. Select a few objects of your choice by selecting, windowing or crossing them and then press [⏎] to continue. Then select Hide Edges and select OK. The selected geometry should have disappeared from the view.

7. Select Edge Properties again and press **U** and [⏎] to unhide all hidden geometry. Select OK to exit the command and the geometry will reappear.

8. Press [⏎] to repeat the Edit Drawing command. Select a point in the exploded isometric view, select Edge Properties and press [⏎] to select geometry. Select a few pieces of geometry of your choice by selecting, windowing or crossing them and press [⏎]. Modify the color and linetype to green and hidden. Select OK twice to exit the command.

9. Issue the Move Drawing View command (AMMOVEVIEW) and practice moving the views around. First select the top view, which is the base view, then try moving the front view and then the isometric views. After practicing moving the views, move the views so that your drawing resembles Figure 8.16.

Figure 8.16

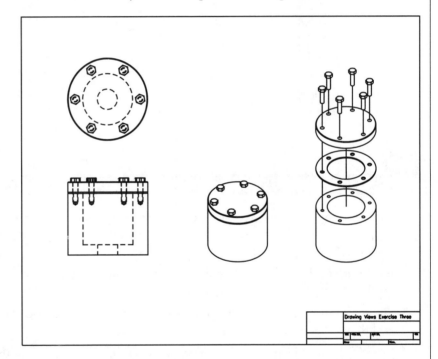

10. Issue the Delete Drawing View command (AMDELVIEW) and select the top assembled view. At the command prompt, press Y to delete all dependent views. All associated views should have been erased.

11. Use the UNDO command to bring back the drawing views, issue the Delete Drawing View command and select the top assembled view. At the command prompt, press **N** to not delete all dependent views. Only the top view should have been erased.

12. Press [Enter] to repeat the Delete Drawing View command and select a point in the isometric view of the assembled parts. The isometric view should have been removed from the screen.
13. Save the file.

Editing Dimensions

The dimensions that automatically appear in the drawing views are parametric dimensions and they can be changed in the same way as you would a parametric dimension on a 2D sketch. Issue the Change Dimension command (AMMODDIM) from the Drawing Layout toolbar, as shown in Figure 8.17, to change a dimension's value, and then update the part with the Update Part command (AMUPDATE) from the Drawing Layout toolbar, as shown in Figure 8.18. The screen will flip back to the part, update it and then automatically switch back to the drawing views and update the views as well.

When drawing views are created, the dimensions do not always appear in the correct location and there may be dimensions that are not required in the views. The dimensions are placed where they were created in the profile. In AutoCAD there are many ways to edit dimensions, and the same is true with Mechanical Desktop. In the following section you will learn how to hide, change the location, align, join, insert, break and modify the text properties.

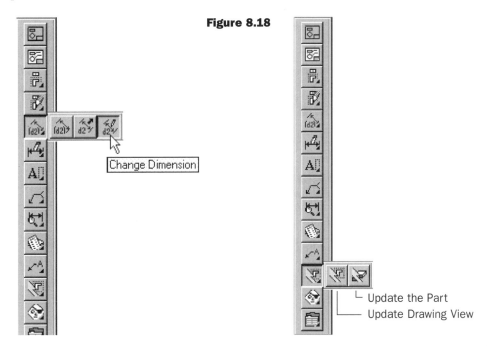

Figure 8.17

Figure 8.18

Hide Dimensions

When drawing views are created, not all the dimensions that appear will be needed for the actual drawing. These dimensions are important to the part, so do not erase them. Instead, hide them in the drawing view. To hide a dimension, issue the Drawing Visibility command (AMVISIBLE) from the Drawing Layout toolbar and a dialog box will appear. From the drawing tab, check Hide and the select Select. You will be returned to the drawing views, where you can select the dimensions to hide. To bring back a hidden dimension, issue the same command, but select Unhide. The dimensions that were hidden will reappear on the screen; select the dimension(s) that you want to unhide.

Figure 8.19

Tutorial 8.4—Hiding Dimensions

1. Open the file \Md2book\Drawings\Drawings\Chapter8\Ex8-4.dwg.

2. Issue the Change Dimension command (AMMODDIM) and change the "1.00" dimension on the right side of the base in the front view to "2.00".

3. Update the part.

4. Update the drawing views if required.

Drawing Views and Annotations

5. Repeat Steps 1 to 3 and change the dimension back to "1.00".

6. Issue the Drawing Visibility command (AMVISIBLE) and select, Hide and Select.

7. In the top view, select the nine "0.00" dimensions.

8. In the right side view, select the two "0.00" dimensions and then select the "3.25" vertical dimension in the same view and press [↵Enter] to return to the dialog box. Select OK to exit the command.

9. Press [↵Enter] to repeat the Drawing Visibility command, select Unhide and Select.

10. Select the "3.25" vertical dimension in the right side view and press [↵Enter] to return to the dialog box. Select OK to exit the command.

11. Save the file.

Move and Reattach Dimensions

If the dimensions are not in the correct location, you have many ways to move the dimensions to a different position. If you are familiar with grips, you can use them in exactly the same way as you would in regular AutoCAD. To use grips while not in any command, select the dimension that you want to relocate with the left mouse button and grip points will appear. Do not select the grips that are located at the points where the dimension meets the geometry. Select the grip on the dimension's text and select a new location. The dimension will move to the new point. To exit grip editing, you can either press [↵Enter] or press the escape key twice. A second method is to use AutoCAD's STRETCH command.

A third method is to use a Mechanical Desktop command called Move Dimension (AMMOVEDIM). This command allows you to flip, reattach or move Mechanical Desktop dimensions within the view or to another view. After issuing the command, press [↵Enter] to move a dimension, select a dimension to move and then select a point in the view for its new location. The view does not need to be the view in which it now appears. However, the view does need to represent the same set of geometry. Then select a new point and the dimension will follow to that point. Continue selecting a point until the dimension is located in the correct position. The other two options are to flip and reattach a dimension. The flip option will change the text of a vertical dimension going from the upper right to the lower left or vice versa. The reattach option is used for dimensions that have their extension lines going through the geometry. Issue the Move Dimension command (AMMOVEDIM), press **R** and [↵Enter]. Select the dimension to reattach and then select the extension line to reattach, select a new location. The same rules apply for reattaching as for creating dimensions. The point will go to the nearest endpoint of a line or center of an arc or circle. If the point that you need to select is under the extension line, you can cycle through the geometry by holding down the control key and selecting the point until the correct object is highlighted. Then press [↵Enter]. Repeat the sequence until the dimensions are correctly reattached.

Tutorial 8.5—Moving Dimensions

1. Open the file \Md2book\Drawings\Drawings\Chapter8\Ex8-5.dwg.

2. Reposition the dimensions using grips, STRETCH or the Move Dimension command (AMMOVEDIM) until your drawing looks like Figure 8.20.

Figure 8.20

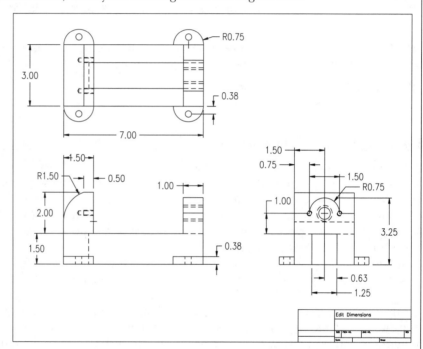

3. Issue the Move Dimension command and press **M** and [enter] to move a dimension. Select the 3/8 vertical dimension in the lower right of the front view. Then select a point in the side view. Your drawing should look like Figure 8.21.

4. Press [enter] to repeat the Move Dimension command and press **R** and [enter] to reattach a dimension. Select the bottom horizontal extension line of the 3/8 dimension that was just moved. Then select near the bottom of the outside vertical line, as shown in Figure 8.22.

5. Select the top horizontal extension line of the same 3/8 dimension. Then select near the top of the outside vertical line, as shown in Figure 8.23. When complete, your drawing should look like Figure 8.24.

6. Save the file.

Figure 8.21

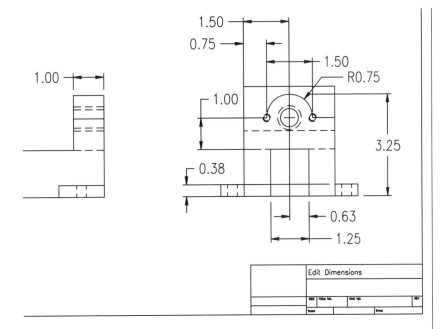

Figure 8.22

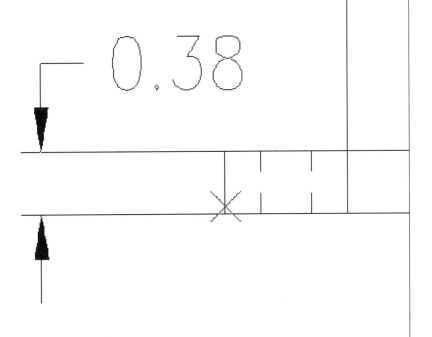

Figure 8.23

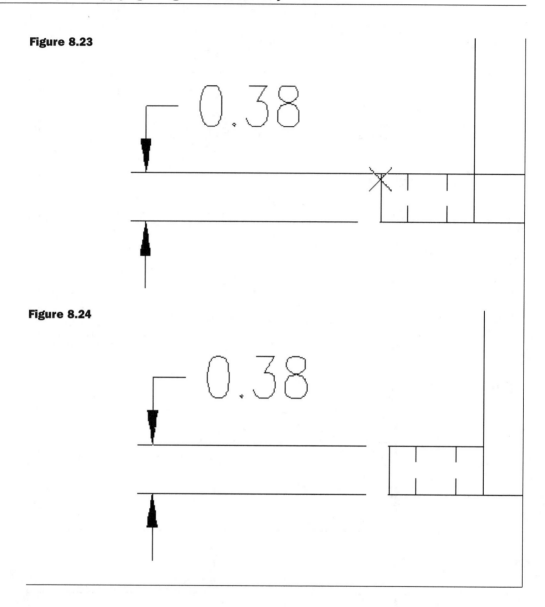

Figure 8.24

Reference Dimensions

After laying out the drawing views, you may find that another dimension is required to better define the part. You could go back and add the dimension to the part if the part was unconstrained by the missing dimension, or you could add a reference dimension. A reference dimension is not a parametric dimension, but it's associative and reflects the length of the geometry being dimensioned. After a reference dimension is created and the views change, the reference dimension will get updated to reflect any changes. A reference dimension is added with the Reference Dimensions command (AMREFDIM) from

Drawing Views and Annotations

the Drawing Layout toolbar, as shown in Figure 8.25 . After issuing the command, create a reference dimension in the same manner as you would a parametric dimension. The reference dimension will be placed on the AMREFDIM layer and will take on the color of that layer. Therefore, if you want to change the color of all the reference dimensions, change the color of the AMREFDIM layer.

Figure 8.25

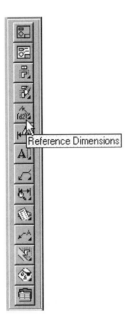

Hole Notes

To create a hole note, issue the Hole Note command (AMHOLENOTE) from the Drawing Layout toolbar, as shown in Figure 8.26. At the command line you will be prompted to edit an existing hole note or to create a new one. Press [↵] to create a new hole note and then select a hole; a dialog box will appear confirming the type of hole. Select the correct type, select OK, position the note and press [↵] to exit the command. The hole note will not reflect the number of holes or tap information; you can add this information by using the same Hole Note command with the Edit option or by using the AutoCAD DDEDIT command.

Figure 8.26

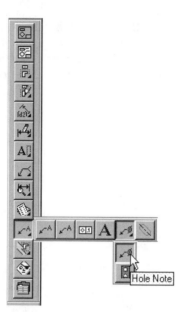

Tutorial 8.6—Adding Reference Dimensions and Hole Notes

1. Open the file \Md2book\Drawings\Drawings\Chapter8\Ex8-5.dwg if it is not the current file.

2. Issue the Reference Dimensions command (AMREFDIM).

3. Place a horizontal and a vertical reference dimension to the four holes in the top view, as shown in Figure 8.27. You may need to relocate "3.00" dimension to make room for the reference dimension.

4. Issue the Hole Note command (AMHOLENOTE) and add three hole notes in the drawing. Then edit them so that your drawing looks like Figure 8.28.

5. Save the file.

Drawing Views and Annotations

Figure 8.27

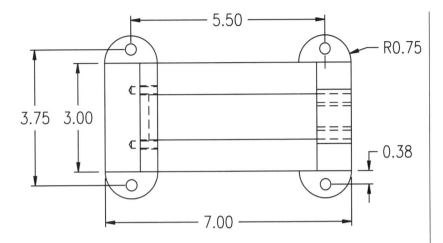

Figure 8.28

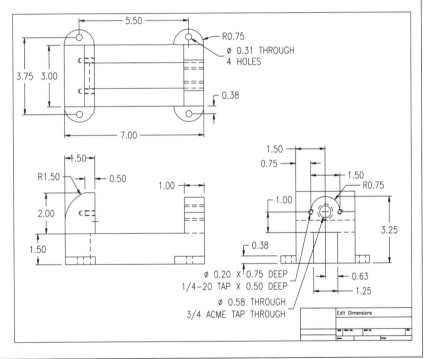

Edit, Align, Join, Insert and Break Dimensions

The dimension style of a dimension that Mechanical Desktop creates can be edited like any other AutoCAD dimension or you can use the Edit Format command (AMEDITFOR-MAT), as shown in Figure 8.29, to modify the format of the dimension. After issuing the command, select the dimension to modify and a dialog box will appear. As you make changes in the dialog box, the change will appear on the dimension. Inside Mechanical Desktop is AutoCAD Mechanical, which adds the capabilities to align, join, insert and break dimensions, as well as several other mechanical drafting aids and standards.

AutoCAD Mechanical commands can be used on Mechanical Desktop dimensions as well as on AutoCAD dimensions. To align dimensions, issue the Align Dimension command (AMDIMALIGN), as shown in Figure 8.29. Select the base dimension that you want the others aligned to, select the dimension(s) that you want to align to the first and press [↵]. The dimensions will be aligned.

The Join Dimensions command (AMDIMJOIN), as shown in Figure 8.29, will join two dimensions into one. The two selected dimensions will be hidden and a reference dimension will be created that joins the two selected dimensions. The reference dimension will be placed at the location of the first dimension selected.

The Insert Dimension command (AMDIMINSERT), as shown in Figure 8.29, will insert a second dimension where there was only one. Issue the command, select a dimension and then select a target location with object snaps for the inserted dimension. The selected dimension will be hidden and two reference dimensions will be created.

In the past, if two dimensions overlapped one another, the only option you had was to explode the dimension and break one of the extension lines. Now with Break Dimension command (AMDIMBREAK), as shown in Figure 8.29, you can break the dimension by selecting two spots where you want it broken. When you break dimensions, it is recommended that you turn off object snaps. The dimension will remain parametric and associative. If the dimension is changed or stretched, the broken area will revert to a solid line.

Figure 8.29

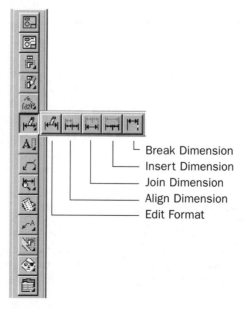

Break Dimension
Insert Dimension
Join Dimension
Align Dimension
Edit Format

Drawing Views and Annotations

Tutorial 8.7—Aligning, Joining, Inserting and Breaking Dimensions

1. Open the file \Md2book\Drawings\Drawings\Chapter8\Ex8-7.dwg.

2. Issue the Align Dimension command (AMDIMALIGN) and select the "0.50" horizontal command in the front view. Then select the "1.00" horizontal dimension in the same view and the "1.50" horizontal dimension in the side view.

3. Press [←Enter] to repeat the Align Dimension command and select the "0.75" horizontal dimension in the side view. Then select the "1.50" horizontal command in the front view and press [←Enter].

4. Press [←Enter] to repeat the Align Dimension command and select the "3.00" vertical dimension in the top view. Then select the "2.00" and "1.50" vertical dimensions in the front view. When you are done aligning dimensions, your drawing should look like Figure 8.30

Figure 8.30

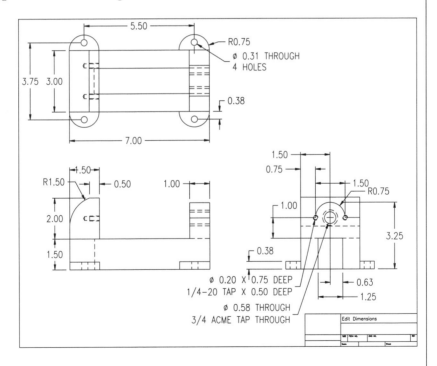

5. Issue the Join Dimensions command (AMDIMJOIN) and select both the "2.00" and "1.50" vertical dimensions in the front view. A "3.50" reference dimension will be created.

Mechanical Desktop 2.0: Applying Designer and Assembly Modules

6. Issue the Insert Dimension command (AMDIMINSERT) and select the "7.00" horizontal dimension in the top view and then select the hole on the bottom left foot.

7. Press [⏎ Enter] to repeat the Insert Dimension command and select the "6.25" horizontal dimension in the top view. Then select the hole in the bottom right foot. You may need to grip edit the "0.75" dimension. When complete, your drawing should resemble Figure 8.31

Figure 8.31

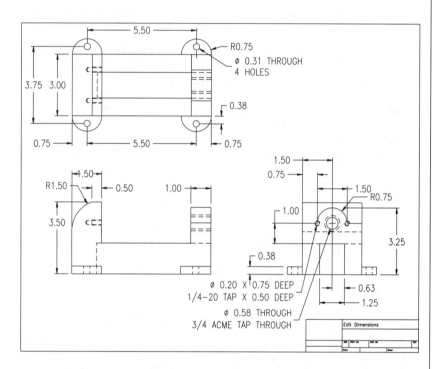

8. Issue the Break Dimension command (AMDIMBREAK) and break the "0.38" dimension in the top view. Press [⏎ Enter] to repeat the Break Dimension command and break the "1.50" dimension in the front view. Press [⏎ Enter] to repeat the Break Dimension command and break one of the "1.50" dimensions in the side view. When complete, your drawing should look like Figure 8.32.

Drawing Views and Annotations

Figure 8.32

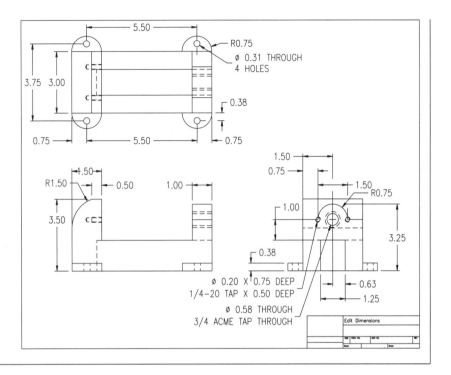

Center Marks, Centerlines and Annotations

To add centerlines to arcs, circles or in between two lines, issue the Centerline command (AMCENLINE) from the Drawing Layout toolbar, as shown in Figure 8.33. Then select an arc or circle and press [⏎] and a center mark will be placed. To place a centerline between two parallel lines, issue the Centerline command, select the two parallel lines, and then select a start and an end point for the centerline. When creating a centerline of a tapped hole, in the side view select the two lines that represent the drill. The lines representing the tap are annotations and cannot be used for centerline placement. These centerlines are attached to the geometry; if the view moves or changes, the centerlines will also move. The distance the centerline goes past the arc or circle can be controlled through Desktop Preferences (under Drawing tab select Centerline).

As you create drawing views, you will find it necessary to add your own 2D geometry, text, etc. to a drawing view. Once the information is in the correct position, you can attach it to a point in a view with the Annotation command (AMMANNOTE) from the Drawing Layout toolbar, as shown in Figure 8.34. Once the information is attached, it will maintain a distance relationship to the selected point. If the view moves or changes, so will the annotation. However, you do not need to add information in an annotation. If the view moves, the information will need to be manually moved. After issuing the Annotation command, press [⏎] to create an annotation, select the geometry or text

Figure 8.33   Figure 8.34

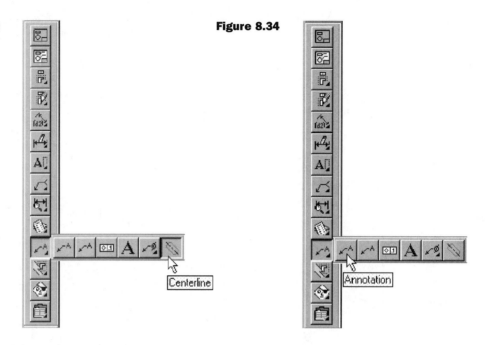

that will be part of the annotation and press [⎯]. Then select a point in a view to attach to. The point selected will go to the nearest endpoint of a line or center of an arc or circle. Once an annotation is created, you can add geometry to it with the Add option. You can delete an annotation with the Delete option and reposition the annotation with the Move option.

Symbols

AutoCAD Mechanical also adds standard symbols such as weld, surface texture, GD&T, datum identifier, datum target, feature identifier and control frames. To create your own standard based on ANSI, BSI, DIN, JIS and ISO, issue the Symbol Standard command (AMSYMSTD) from the Drawing Layout toolbar, as shown in Figure 8.35. You can copy a standard and then customize the copied standard to meet your requirements. The method for creating all the symbols is the same: issue the command, select a starting point and fill in the information. As the information in the dialog box is filled in, it will also appear on the screen. The symbols can be used in Mechanical Desktop views as well as in AutoCAD 2D drawings.

To attach a symbol to a view, select the attach button in the dialog box and then select a point in the view to attach it to. If there is not an Attach option in the first dialog box, you will find it under the Leader tab. To edit an existing symbol, issue the AMEDIT command from the Drawing>Annotate>Edit Symbols... pull-down menu or type it in and select the symbol. The same dialog box where it was created will appear.

Figure 8.35

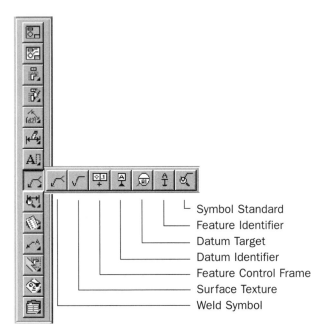

Tutorial 8.8—Center Marks, Center Lines and Symbols

1. Open the file \Md2book\Drawings\Chapter8\Ex8-8.dwg.

2. Issue the Centerline command (AMCENLINE) and add all the centerlines as shown in Figure 8.36.

3. Add two surface finish symbols to the front view; use the attach option from the Leader tab.

4. Move a few of the views around to verify that the center marks, centerlines and surface finish symbols also move with the view.

5. Save the file.

Figure 8.36

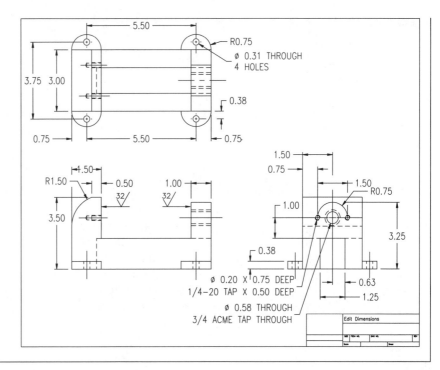

Creating a Bill of Materials

The final step in a drawing is to create a bill of materials. First issue the Create a BOM Table command (AMBOMSETUP) from the Drawing Layout toolbar, as shown in Figure 8.37, to set the appearance of the balloons as well as the bill of materials table. Before you create a balloon, the appearance needs to be set, as well as the type of information it should contain. Figure 8.38 shows the dialog box for setting up the bill of materials.

In the **Table definition** section:

Parts/Subassemblies to Include: Select whether to include in the bill of materials table all parts and subassemblies or only those that have been ballooned.

Justification: Text justifies to the left or right.

Sort Column: Select the column name in the Defined Column Names area and then select Sort Column. The bill of materials table will be sorted by this column.

Sort Order: Sort the bill of materials up or down or do not sort.

Column Order: Move a column name up or down from its current position.

Drawing Views and Annotations

Figure 8.37

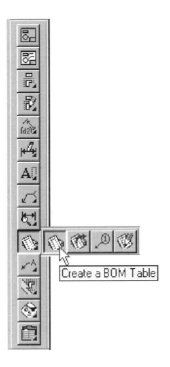

Figure 8.38

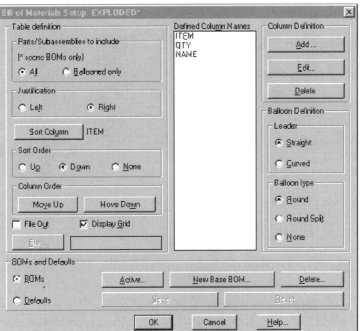

File Out: Select whether you want to create an external file.

Grid Display: If checked, creates a grid of lines around the bill of materials.

In the **BOMs and Defaults** section, select BOMs or Defaults.

For BOMs:

Active...: Select a bill of materials to make active.

New Base BOM...: Create a new bill of materials in the same drawing.

Delete...: Delete an existing bill of materials.

For Defaults:

Save: After making changes, select Default and then Save. These new values are now the default.

Reset: Resets to the original settings.

The **Defined Column Names** section lists the column names for the current bill of materials.

In the **Column Definition** section:

Add: Adds a column to the bill of materials. In a dialog box you will fill in a name, width (number of characters), attribute, column type and justification. A width of "0" will be "autofit" to fit the number of characters.

Edit: Edits a column.

Delete: Deletes a column.

In the **Balloon Definition** section:

Leader: Specify whether the balloon leader will be straight or curved.

Balloon Type: Specify whether the balloon is round, round split, or none.

Balloons

After the bill of materials information is set, create the balloons by issuing the Create Balloon command (AMBALLOON) from the Drawing Layout toolbar, as shown in Figure 8.39 . At the command prompt, press [⏎] to dimension a part and then select a part

Drawing Views and Annotations

to balloon. Then select a point where the leader will start and keep selecting points that will add segments to the leader line. Press [Enter] when done adding segments. A dialog box will prompt you to confirm the information by selecting a column name. To change the information, type in the new information over the old. The default item number reflects the order that it is assigned in the browser. An item number cannot be changed to a number of an existing part. The name is the part name that it was given when it was created. Use grips to edit a location of a balloon leader.

Any changes made to the part, and the addition or removal of instances, will automatically be updated in the balloon as well as in the bill of materials table. To edit the content of a balloon, issue the Edit BOM Entry command (AMEDITBOM) from the Drawing Layout toolbar, as shown in Figure 8.40, and a dialog box will appear, as shown in Figure 8.41.

Figure 8.39

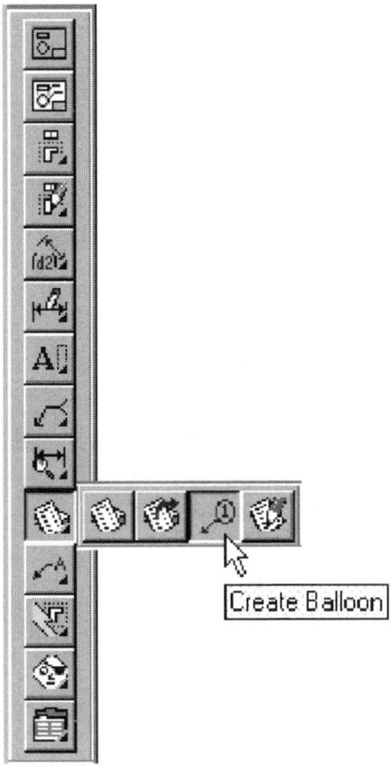

Figure 8.40

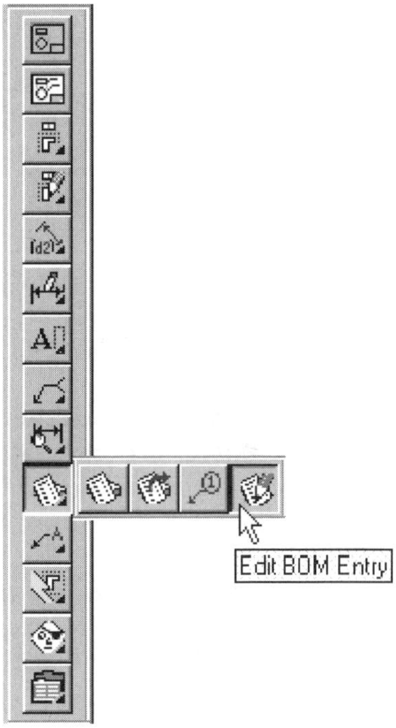

Edit: After selecting a balloon number, select Edit to modify the contents on the balloon.

Delete: Select a balloon number to delete.

Hide: Select a balloon number to hide in the bill of materials.

Lock: Select a balloon number and then the bill of materials will not automatically update when changes are made to the part.

Figure 8.41

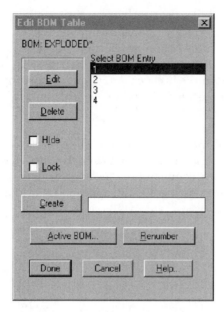

Create: Creates an item number without having to create a balloon.

Active BOM: Select a bill of materials to make active.

Select BOM Entry: Select a balloon number to change.

Renumber: After an item number is edited or hidden, the balloons as well as an existing bill of materials table will be renumbered sequentially starting at 1. Hidden numbers are given large numbers like 100001, 100002, etc. and they will not show in the bill of materials.

To create the bill of materials, issue the Insert BOM Table into Drawing command (AMBOM) from the Drawing Layout toolbar, as shown in Figure 8.42. You will be prompted for three points. The first two points define the extents of the bill of materials and the third point defines the insertion point. If the first two points were selected bottom to top, the insertion point will be the lower right corner. If the first two points were selected top to bottom, the insertion point will be the upper right corner. The text height is controlled by the dimension text height of the current dimension style. If the bill of materials is longer than the specified height, a second column will be created next to the first bill of materials; it will wrap. To move a bill of materials, you can use grips or the Move command. Any edits done to the bill of materials will be reflected in the bill of materials.

Figure 8.42

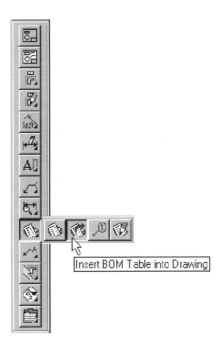

Tutorial 8.9—Creating a Bill of Materials

1. Open the file `\Md2book\Drawings\Chapter8\Ex8-9.dwg`.

2. Switch to the drawing views if they are not current.

3. Issue the Create a BOM Table command (AMBOMSETUP) and change the Sort Order to up.

4. Issue the Create Balloon command (AMBALLOON) and balloon the exploded view, as shown in Figure 8.43.

5. Make the scene "Exploded" your active scene.

6. Return to the drawing mode.

7. Issue the Insert BOM Table into Drawing command (AMBOM) and create a bill of materials as shown in Figure 8.43.

8. Save the file.

Figure 8.43

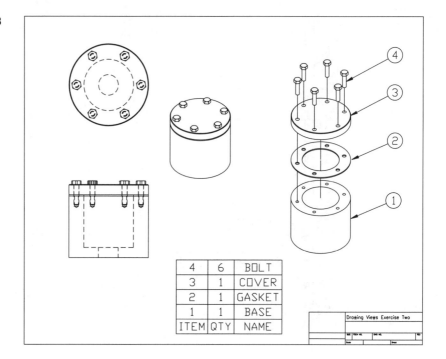

Exercises

For the following exercises, follow the instructions before each.

Exercise 8.1—Drawing Views with Annotations

Open the file \Md2book\Drawings\Chapter8\Foot.dwg. Create drawing views with a scale of 1/2, reposition dimensions, add reference dimensions and a hole note, as shown in Figure 8.44, and then save the file.

Exercise 8.2—Drawing Views with Balloons and a Bill of Materials

Open the file \Md2book\Drawings\Chapter8\Guide.dwg. Create drawing views with a scale of 1/2 from the scenes Assembled and Exploded. Then add balloons and a bill of materials table, as shown in Figure 8.45.

Figure 8.44

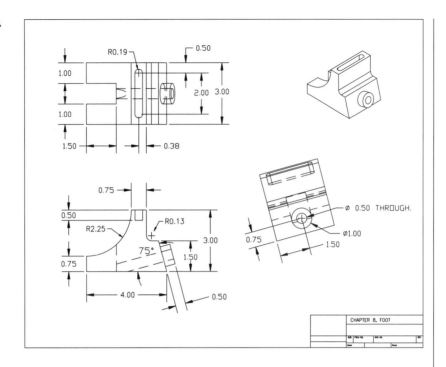

Figure 8.45

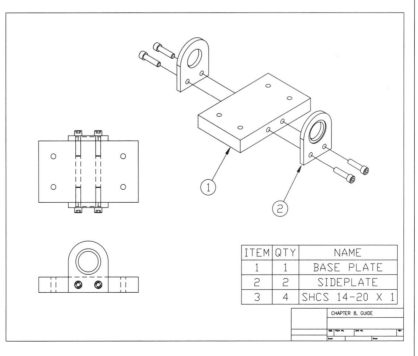

Review Questions

1. There can be only one base view in a drawing. T or F?

2. The color and linetype of layers that Mechanical Desktop information is placed on cannot be changed. T or F?

3. Why is it not recommended to use the Select option when creating drawing views?

4. The only way to create an isometric view is to first create a base view and then create an isometric view from it. T or F?

5. When plotting a drawing with Mechanical Desktop views, you always plot at 1=1. T or F?

6. Explain how a dimension can be moved from one view to another and also how to reattach a dimension.

7. Reference dimensions can be changed to modify the part. T of F?

8. Explain what an annotation is in reference to Mechanical Desktop drawing views.

9. The functionality of AutoCAD Mechanical cannot be used with Mechanical Desktop drawings. T or F?

10. A bill of materials entry is fully associative to the part as long as it is not locked. T of F?

chapter 9

Practice Exercises

In this chapter you will be guided through three exercises, intended to demonstrate different modeling and assembling techniques, not design principles. Through the exercises, you will apply the lessons that were learned in the previous chapters. Follow the steps to create the parts and assemble the parts. While working on the exercises, use any of the viewing techniques to better visualize the parts. When the exercises are complete, go back and experiment with other techniques in addition to creating drawing views.

In the directory `\Md2book\Drawings\Chapter9\Finished` you will find the completed exercises for the Dryer, Car Stand and Wood Plane assembly. If you need help with a particular part, open the completed file and use the Feature Replay command (AMREPLAY) to sequence through to see how the part was built.

Exercise 9.1—Hair Dryer

In this exercise you will create a hair dryer (Figure 9.1).

1. Start a new drawing.
2. Sketch, profile and dimension the geometry as shown in Figure 9.2.
3. Change to an isometric view (**8**) and press ⏎.
4. Revolve the profile 180° about the bottom horizontal line and flip the direction down. When complete, your drawing should look like Figure 9.3.
5. Change back to the plan view (**9**) and press ⏎.
6. Sketch, profile and dimension the rectangle as shown in Figure 9.4.

Figure 9.1
Completed
Hair Dryer

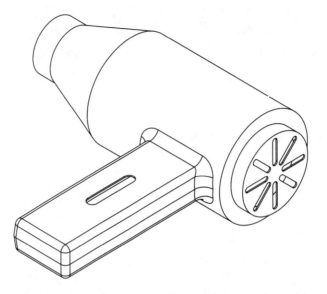

Figure 9.2

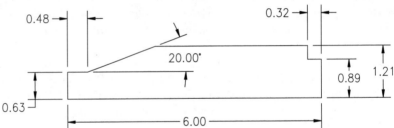

Figure 9.3

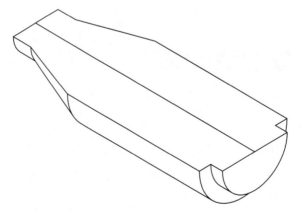

7. Change to an isometric view (**8**) and press ⏎.

8. Extrude and join the rectangle ".375" with a draft angle of "-5" and flip the direction down. When complete, your drawing should look like Figure 9.5.

Figure 9.4

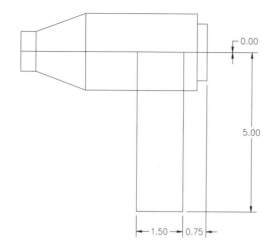

Figure 9.5

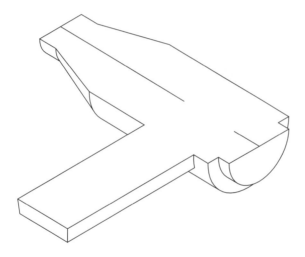

9. Add ".125" fillets to the five highlighted edges, as shown in Figure 9.6.

Figure 9.6

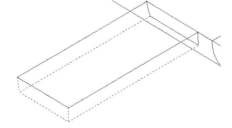

10. Shell the part ".03" and exclude the top face and the front arc that defines the front of the dryer. When complete, your drawing should look like Figure 9.7.

Figure 9.7

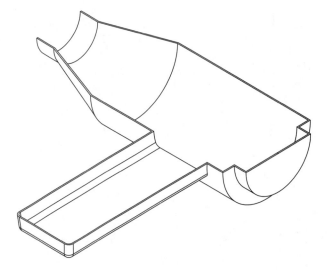

11. To create the top half of the dryer, in the same file, mirror the bottom half about the plane that is highlighted in Figure 9.8 and press [Enter] to create a new part. When complete, your drawing should look like Figure 9.9.

Figure 9.8

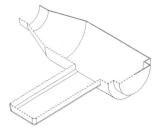

Figure 9.9

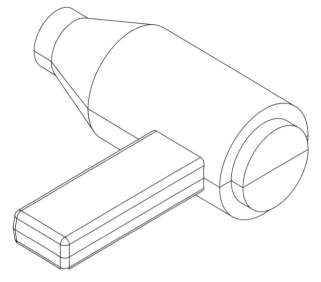

12. There are now two individual parts on your screen. Combine both parts with the Join option. Select the bottom half to join to the top half, because the top half is the active part. When complete, you should have a single part. You can verify this in the browser by expanding the combine feature in the browser to see Part1_1, as shown in Figure 9.10.

Figure 9.10

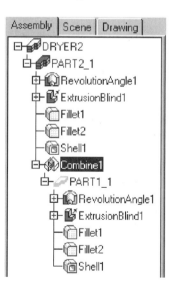

13. Add a ".2" fillet to the highlighted edge, as shown in Figure 9.11. This will create one fillet going all the way around the dryer.

Figure 9.11

14. Make the back of the dryer (the circle that is highlighted in Figure 9.12) the active sketch plane and orient the UCS by selecting one of the long vertical lines that define the handle. When complete, your drawing should look like Figure 9.12; the circle should not be highlighted when done.

Figure 9.12

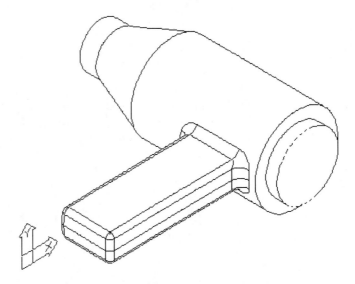

15. Sketch, profile and dimension the slot, as shown in Figure 9.13. Add an XValue constraint to an arc on the slot and the outside circle that defines the back of the dryer.

Figure 9.13

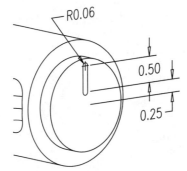

Practice Exercises

16. Extrude the slot with no draft angle, Cut and To Plane option selected. For the plane, select the inside of the shell as shown in Figure 9.14.

Figure 9.14

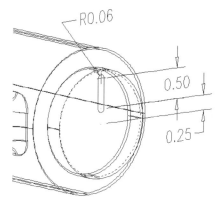

17. Before arraying the slot, place a work axis through the center of the part and then do a polar array with "8" instances, Full circle, Rotate as copied and then array the slot around the work axis. When complete, your drawing should look like Figure 9.15.

Figure 9.15

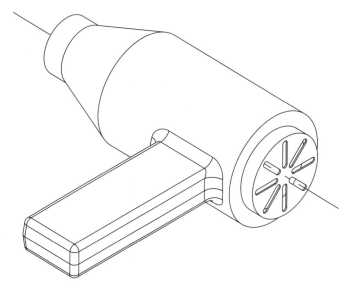

18. Make the top plane of the handle (highlighted in Figure 9.16) the active sketch plane and orient the UCS by selecting one of the edges that define the handle. When complete, your drawing should look like Figure 9.16; the plane will not be highlighted when complete.

Figure 9.16

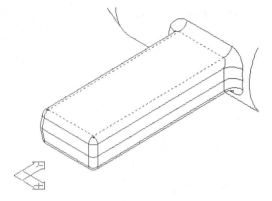

19. Sketch, profile and dimension the slot, as shown in Figure 9.17.

Figure 9.17

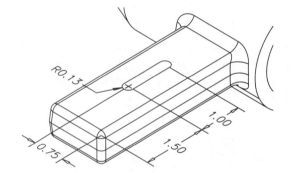

20. Extrude slot with the Cut and To Plane option. For the plane, select the inside of the shell, as shown in Figure 9.18.

Figure 9.18

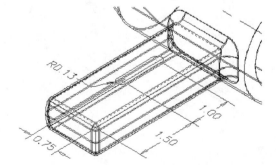

21. Turn off work axis. When complete, your part should look like Figure 9.19.

Practice Exercises

Figure 9.19

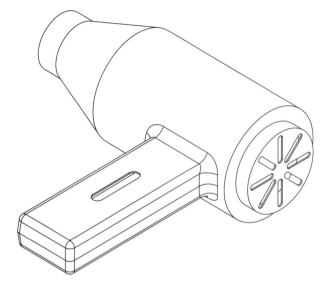

22. Save the file as \Md2book\Drawings\Chapter9**Dryer.dwg**.

Exercise 9.2—Car Stand

In this exercise, you will create a car stand that consists of three parts in a top down assembly (all the parts will exist in the same file).

1. Start a new drawing.
2. Sketch, profile and dimension a "2" square.
3. Rename the part to Base Stand.
4. Change to an isometric view (**8**) and press ⏎.
5. Extrude the square "7" with a "20" draft angle and flip the extrusion direction so that it is going down into the screen. When complete, your drawing should resemble Figure 9.21.
6. Create a linear fillet on each of the angled edges with a radius of "1" at the bottom and ".5" at the top of the extrusion. When complete, your drawing should look like Figure 9.22.
7. Shell out the part with a thickness of ".125" and exclude the bottom of the extrusion. When complete, your drawing should look like Figure 9.23.

Figure 9.20
Completed
Car Stand

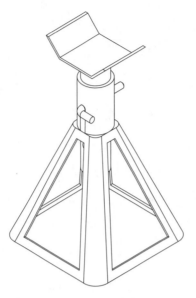

Figure 9.21

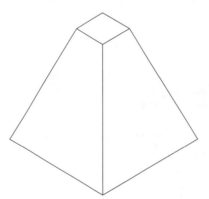

Figure 9.22

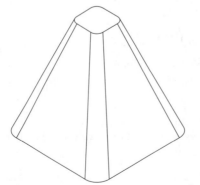

Practice Exercises

Figure 9.23

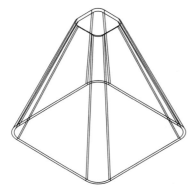

8. Draw a circle on the top of the extrusion, profile it and dimension it as shown in Figure 9.24. Since the top of the extrusion is still the active sketch plane, you do not need to make it the active sketch plane.

Figure 9.24

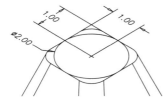

9. Extrude the circle "3", no draft angle and join it to the extrusion, accepting the default direction out of the part.
10. Create a "1.625" diameter drilled hole concentric through the cylinder. When complete, your drawing should look like Figure 9.25.

Figure 9.25

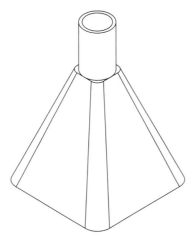

11. Create a work axis through the cylinder.
12. Create a work plane On Edge - Planar Parallel and check Create Sketch Plane, select the work axis and then type in "ZX". Orient the UCS by selecting the work axis. When complete, your drawing should look like Figure 9.26.

Figure 9.26

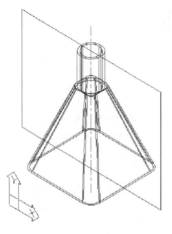

13. Change to the active sketch view (**9**) and press [⏎].
14. Turn off the visibility of all work planes.
15. Draw a circle near the middle of the cylinder, profile it and dimension it as shown in Figure 9.27. Add an XValue constraint to the circle and a circle that defines the cylinder. This will fully constrain the part. A work point could also have been created on the outside of the cylinder to place in this feature. When the holder is created, you will place a hole with the work point option.

Figure 9.27

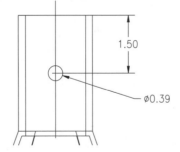

16. Change to an isometric view (**8**) and press [⏎].
17. Extrude the circle Mid Plane with the Cut option and a distance of 2.

18. Make the front left face of the extrusion (the face highlighted in Figure 9.28) the active sketch plane and orient the UCS using the bottom horizontal line defining the plane. When complete, your drawing should look like Figure 9.28; the face should not be highlighted when done.

Figure 9.28

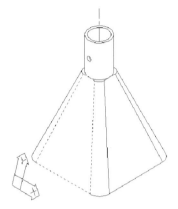

19. Sketch, profile and dimension the geometry as shown in Figure 9.29.

Figure 9.29

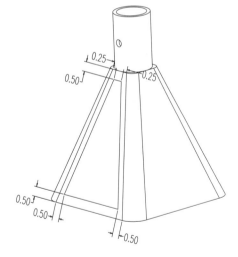

20. Extrude the profile ".25" to remove material from the side of the first extrusion.

21. Make the bottom of the base extrusion the active sketch plane and orient the UCS to one of the edges defining the bottom plane.

22. Array the cutout around the work axis with Polar Array, 4 instances, Full circle and Rotate as copied.

23. Turn the work axis off. When complete, your drawing should look like Figure 9-30.

Figure 9.30

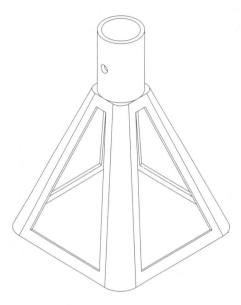

24. Save the file as \Md2book\Drawings\Chapter9\CarStand.dwg.
25. Create a new part named **PIN** in the same file.
26. For the pin you will sweep a circle around a path. Sketch the path by issuing the Path command (AMPATH) and select the end of the ".75" line as the starting point. Then dimension the path as shown in Figure 9.31.

Figure 9.31

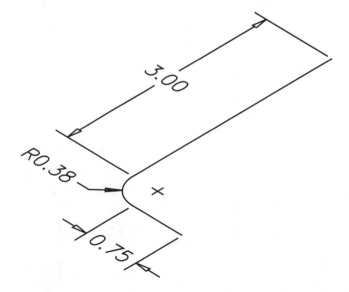

Practice Exercises

27. Create a work plane with the Sweep Profile option and make it the Current Sketch Plane. Rotate the UCS until the X is pointing into the screen.

28. Draw a circle near the middle of the work point, profile it and add a ".375" diameter dimension, as shown in Figure 9.32. Add a concentric constraint to the circle and the work point. This will fully constrain the part.

Figure 9.32

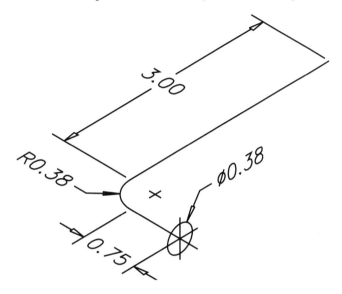

29. Sweep the circle along the path and accept the defaults in the dialog box.

30. Add a "0.0625" chamfer with the Equal Distance option to the circular edge at the end of the 3"-long sweep.

31. Turn the visibility of the work planes off. When complete, your drawing should resemble Figure 9.33.

Figure 9.33

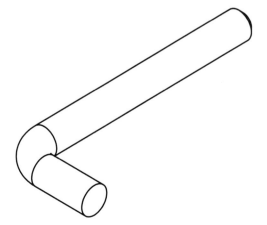

32. Save the file.
33. Switch to the world coordinate system by typing in "UCS" and pressing ⏎ twice.
34. Create a new part named HOLDER in the same file.
35. Draw a circle, profile it, and dimension it with "1.625" diameter dimension.
36. Extrude the circle "7" and flip the extrusion direction so that it is going down into the screen.
37. Create a "1.375" concentric drilled hole through the cylinder.
38. Place a work axis through the cylinder.
39. Create a work plane using On Edge/Axis - Planar Parallel and Create Sketch Plane, select the work axis for the edge and type "ZX" as the plane to be parallel to. Select the work axis to orient the UCS and rotate the UCS once until the X is pointing toward the command line.
40. Turn off the visibility of the work plane.
41. Draw, profile and dimension the geometry as shown in Figure 9.34.

Figure 9.34

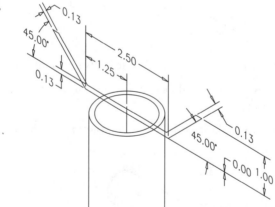

42. Extrude the profile "3" Mid Plane and join it to the holder.
43. Turn the visibility of work plane back on.
44. Create a work plane that is Tangent - Planar Parallel and make the work plane the active sketch plane. Select near the southwest quadrant of the cylinder and select the existing work plane to be parallel to. Select the work axis to orient the UCS until the X is pointing toward the command line.
45. Turn off all work planes and work axis. Then zoom into the top of the holder.

Practice Exercises

46. Create a work point and dimension it as shown in Figure 9.35.

Figure 9.35

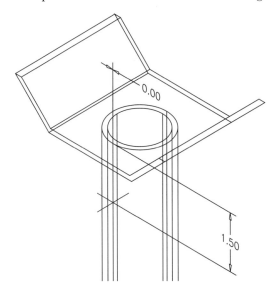

47. Create a ".38" diameter drill hole through the cylinder using the On Point option.

48. Array the drilled hole with "4" Rows and a Spacing of "-1.25". When complete, your drawing should look like Figure 9.36.

Figure 9.36

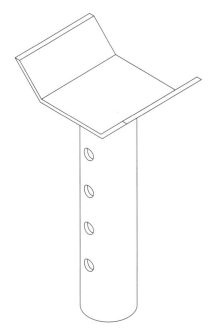

49. Save the file.
50. Before assembling the three parts, move the holder near the Base Stand.
51. Use the Mate constraint to align the center of the Base Stand and the Holder. Figure 9.37 shows the two centerlines of the parts. Press [↵Enter] to accept the default offset distance of zero.

Figure 9.37

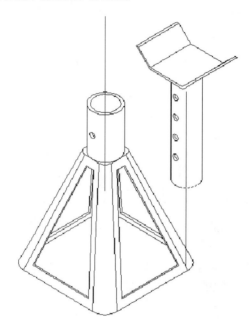

52. Press [↵Enter] to repeat the Mate constraint and align the center of the hole in the Base Stand and the hole that is second from the top in the Holder. Figure 9.38 shows the two centerlines of the parts (the centerline of the holder does appear off the part). Press [↵Enter] to accept the default offset distance of zero.
53. Move the Pin near to the Base Stand.
54. Use the Mate constraint to align the center of the Base Stand and Pin. Figure 9.39 shows the two centerlines of the parts. Press [↵Enter] to accept the default offset distance of zero.
55. Press [↵Enter] to repeat the Mate constraint and select the inside quadrant of the pin, using the quadrant object snap, as shown in Figure 9.40. Cycle through the constraints until the arrow points inward (this represents the tangent face at the inside quadrant). Press [↵Enter] to accept this option. Select the front quadrant of the bottom of the cylinder of the Base Stand, using the quadrant object snap as shown in Figure 9.41. Cycle through the constraints until the arrow points outward and press [↵Enter] to accept this option. At this point, your

Practice Exercises

Figure 9.38

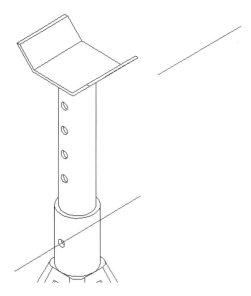

Figure 9.39

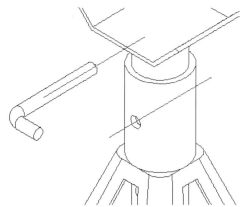

drawing should look like Figure 9.42. When complete, your drawing should look like Figure 9.43. If the pin is on the opposite side, update the assembly to switch the pin to the correct side.

56. Save the file.

Figure 9.40

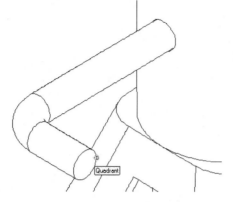

Figure 9.41

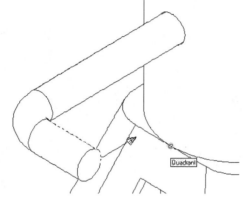

Figure 9.42

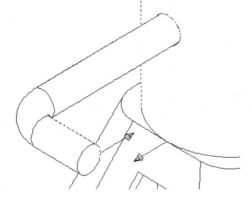

Figure 9.43

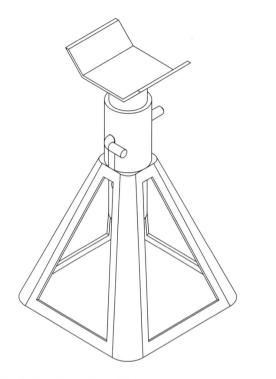

 Exercise 9.3—Wood Plane

In this exercise, you will create a wood plane that consists of five parts in a bottom up assembly (each part is in its own file). You will create four of the parts in their own files. The fifth part is a flat head screw, which can be found on your CD: \Md2book\Drawings\Chapter9\10-32x1.dwg.

1. Start a new drawing for the base of the wood plane.
2. Sketch, profile and dimension the geometry as shown in Figure 9.45. The two construction lines should have tangent constraints applied to them and the arc they touch.
3. Switch to an isometric view (**8**) and press ⌐⌐.
4. Extrude the profile "1.875" in the default direction.
5. Use the ROTATE3D command to rotate the part 90° about the X axis.
6. Make the right side face the active sketch plane and orient the UCS as shown in Figure 9.46.

Figure 9.44
Completed
Wood Plane

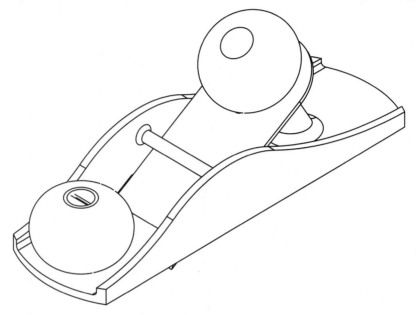

Figure 9.45

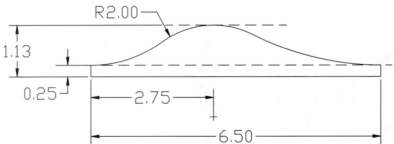

Figure 9.46

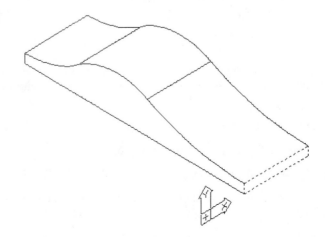

Practice Exercises

7. Sketch, profile and dimension the rectangle as shown in Figure 9.47.

Figure 9.47

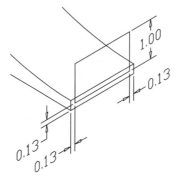

8. Extrude the rectangle through the part, cutting away the inside.
9. Make the bottom plane the active sketch plane and orient the UCS as shown in Figure 9.48. You will need to do a Z flip.

Figure 9.48

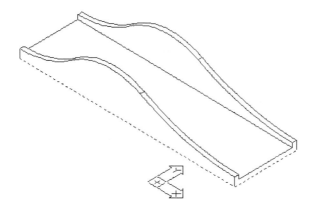

10. Change to sketch view (**9**) and press ⬜.
11. Sketch and profile the geometry as shown in Figure 9.49.

Figure 9.49

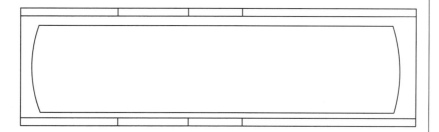

12. Add two "0" dimensions to the ends of the lines, add a "6" dimension to one of the lines and add two tangent constraints to the arcs and outside vertical edges. Add two collinear constraints to the lines and outside horizontal edges and add a radius constraint to both arcs to fully constrain the profile. When complete, your drawing should look like Figure 9.50.

Figure 9.50

13. Change to an isometric view (**8**) and press [⏎].
14. Extrude the profile through the part using the Intersect option. When complete, your drawing should look like Figure 9.51.

Figure 9.51

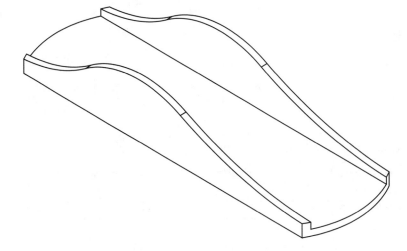

15. Next make the inside plane the active sketch plane and orient the UCS as shown in Figure 9.52.
16. Draw two vertical lines close to the inside edges, profile them and select the two inside edges to close the profile. Then add the two dimensions as shown in Figure 9.53.

Figure 9.52

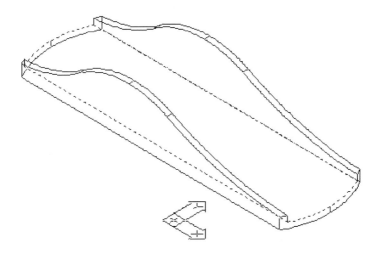

Figure 9.53

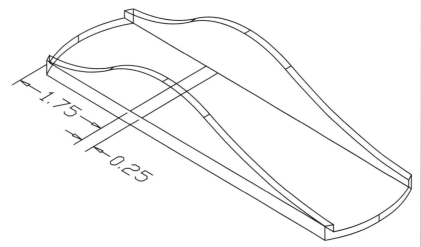

17. Extrude the profile with the Cut and Through options and remove the material through the bottom of the part.

18. Create a chamfer with the Distance x Angle option, setting the distance to ".125" and the angle to "60". Select the edge that is highlighted in Figure 9.54. For the prompt:

    ```
    Apply angle value to highlighted face:
    ```

    press **N** and [⏎] to highlight the inside vertical face and then press [⏎] to accept this face.

19. Change to a southwest isometric view (**88**) and press [⏎].

Figure 9.54

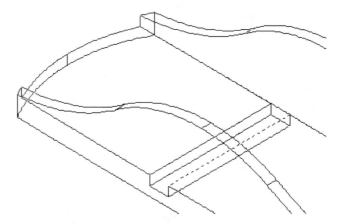

20. Make the chamfered face the current sketch plane and orient the UCS about the top edge of the chamfer. Rotate the UCS until your drawing looks like Figure 9.55.

Figure 9.55

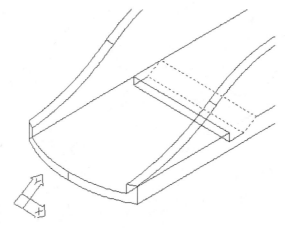

21. Change to the plan view (**9**) and press [←Enter].

22. Draw a circle, profile it and dimension it as shown in Figure 9.56. Add an XValue constraint to the circle and one of the bottom arcs in the middle of the base of the plane; this will center the circle in the part.

23. Extrude the circle using the Join and To Plane option, selecting the inside plane. When complete, your drawing should resemble Figure 9.57.

24. Make the outside vertical face the active sketch plane, as highlighted in Figure 9.58, and orient the UCS as shown in Figure 9.58.

Figure 9.56

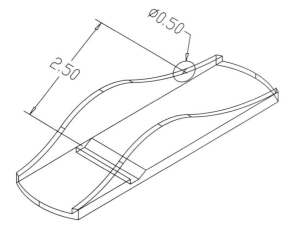

Figure 9.57

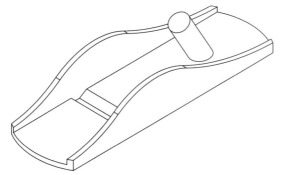

Figure 9.58

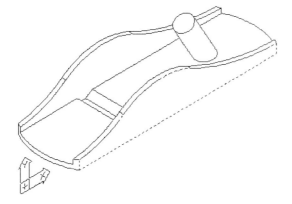

25. Draw a circle, profile it and dimension it as shown in Figure 9.59. Add an XValue constraint to the circle and one of the top arcs on the side of the plane.
26. Extrude the circle using the Join and To Plane option and select the opposite outside plane. When complete, your drawing should look like Figure 9.60.

Figure 9.59

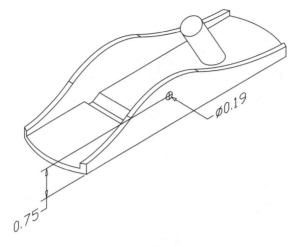

Figure 9.60

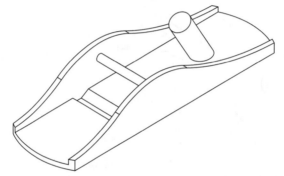

27. Make the inside face the active sketch plane, as highlighted in Figure 9.61, and orient the UCS as shown in Figure 9.61.

Figure 9.61

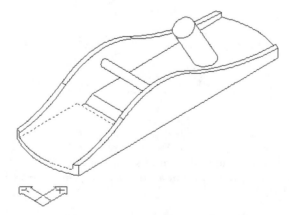

28. Create a work point and dimension it as shown in Figure 9.62.

Practice Exercises

Figure 9.62

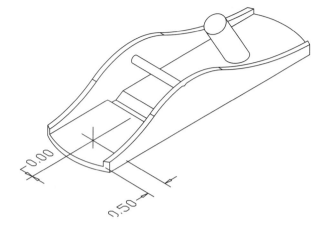

29. Create a #10-32 through tapped hole on the work point (.159 diameter for the drilled through hole and .19 major diameter for the tap).

30. Create a #10-32 x 1 blind tapped hole concentric to the angle extrusion (.159 diameter for the drilled through hole and .19 major diameter for the tap).

31. Add six ".06" constant fillets to the inside edges of the plane and the circular cross bar. You may need to rotate the part to clearly see all six edges.

32. Add a ".1875" fixed width fillet to the bottom of the angled circular extrusion. When complete, your drawing should look like Figure 9.63.

Figure 9.63

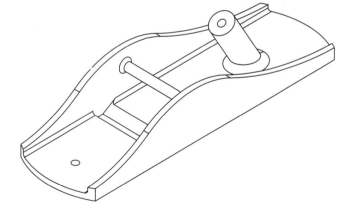

33. Save the file as \Md2book\Drawings\Chapter9\WDPlane.dwg.

34. Start a new drawing for the blade for the wood plane.

35. Sketch, profile and dimension the geometry as shown in Figure 9.64.

36. Switch to an isometric view (**8**) and press [⏎].

Figure 9.64

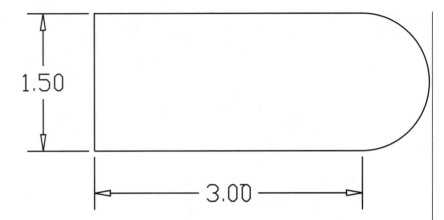

37. Extrude the profile ".0625" in the default direction.
38. Sketch, profile and dimension the slot as shown in Figure 9.65. Add a concentric constraint to the arc on the front of the slot and the arc on the blade.

Figure 9.65

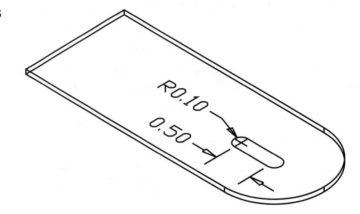

39. Extrude the profile with the Cut and Through options.
40. Create a chamfer with the Distance x Angle option, setting the distance to ".0625" and the angle to "60". Select the edge that is highlighted in Figure 9.66. Press [⏎] to accept the back face at the prompt:

    Apply angle value to highlighted face.

    When complete, your drawing should look like Figure 9.67.

Practice Exercises

Figure 9.66

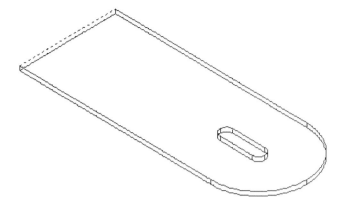

Figure 9.67

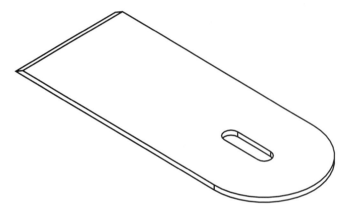

41. Save the file as `\Md2book\Drawings\Chapter9\Blade.dwg`.
42. Start a new drawing for the front handle of the wood plane.
43. Sketch, profile and dimension the geometry as shown in Figure 9.68.

Figure 9.68

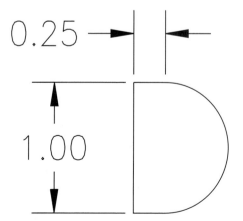

44. Switch to an isometric view (**8**) and press [⏎].
45. Do a Full revolution around the vertical line.
46. Create a counter sink hole for a #10-32 flat head screw that is concentric to the top of the handle, (drill size Dia = ".2", C'Dia = ".4" and C'Angle = 45). When complete, your drawing should look like Figure 9.69.

Figure 9.69

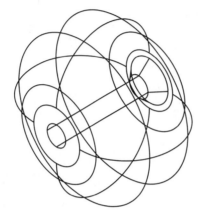

47. Save the file as \Md2book\Drawings\Chapter9\FrntHndl.dwg.
48. Using Windows Explorer, copy the file

    \Md2book\Drawings\Chapter9\FrntHndl.dwg

    to a new file named

    \Md2book\Drawings\Chapter9\BldHndl.dwg.

49. Open the file \Md2book\Drawings\Chapter9\BldHndl.dwg. In this file you will create a subassembly of the blade handle and a threaded rod.
50. Delete the countersink hole in the handle.
51. Create a #10-32 x .5 blind tapped hole concentric to the bottom of the handle (.159 diameter for the drilled through hole, .19 major diameter for the tap and a point angle of 118).
52. Create a new part named **THREAD** in the same file. You will create a cylinder that will represent a threaded rod.
53. Draw a circle, profile it and dimension it with a ".19" diameter.
54. Extrude it "1" in the default direction.
55. Add a ".03" equal distance chamfer to both circular edges of the cylinder. When complete, your drawing should resemble Figure 9.70.

Figure 9.70

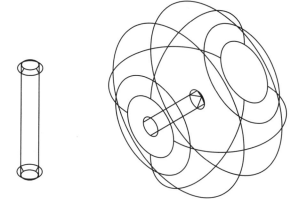

56. Use the Mate constraint to align the center of the blade handle and the thread. Figure 9.71 shows the two centerlines of the parts. Press [⏎] to accept the default offset distance of zero.

Figure 9.71

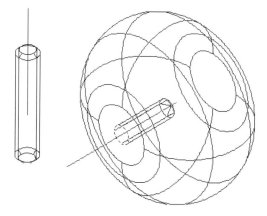

57. To better select the geometry, move the parts so that they are not touching one another. Use the Mate constraint to align the center points of the inside of the blade handle and the top of the thread. Figure 9.72 shows the two center points of the parts. Press [⏎] to accept the default offset distance of zero. When complete, your drawing should resemble Figure 9.73.

58. Save the file.

59. Start a new drawing that will become the assembly file for the wood plane assembly.

60. Switch to an isometric view (8) and press [⏎].

61. From the Assembly Catalog dialog box, add the directory `\Md2book\Drawings\Chapter9`.

Figure 9.72

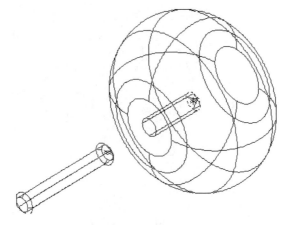

Figure 9.73

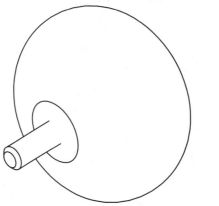

62. Attach one instance of each of the files Plane, FrntHndl, 10-32x1, Blade and BldHndl. When complete, your drawing should resemble Figure 9.74.

Figure 9.74

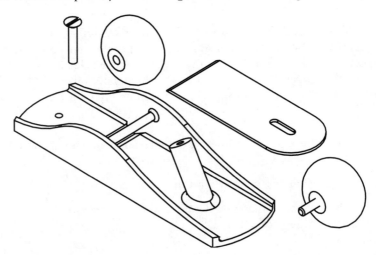

Practice Exercises

63. Use the Insert constraint to align the bottom of the front handle and the front hole. Figure 9.75 shows the two centerlines of the parts. Press [←Enter] to accept the default offset distance of zero.

Figure 9.75

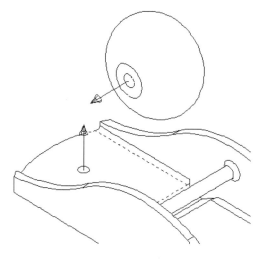

64. Use the Mate constraint to align the center of the screw and the front handle. Figure 9.76 shows the two centerlines and faces of the parts. Press [←Enter] to accept the default offset distance of zero.

Figure 9.76

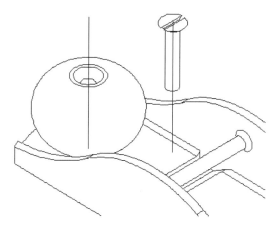

65. To better select the geometry, move the parts so that they are not touching one another. Use the Mate constraint to align the center points of the top of the threads of the screw and the top of the drill diameter of the hole. Figure 9.77 shows the two center points of the parts. Press [←Enter] to accept the default offset distance of zero.

Figure 9.77

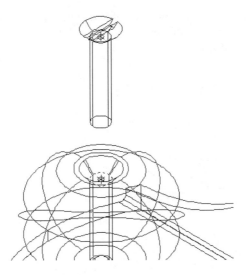

66. Change to a southwest isometric view (**88**) and press [↵Enter].

67. Use the Mate constraint to mate the bottom of the blade and the top of the chamfer. Figure 9.78 shows the two mating faces of the parts. Press [↵Enter] to accept the default offset distance of zero.

Figure 9.78

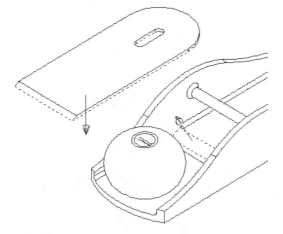

68. To position the blade from side to side, press [↵Enter] to repeat the Mate constraint. Select the outside edge of the blade and the inside edge of the plane as shown in Figure 9.79. Type in an offset distance of ".0625".

Figure 9.79

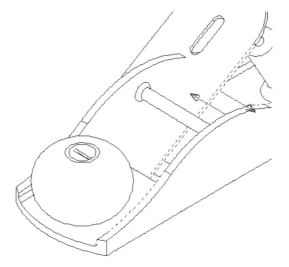

69. To position the slot in the blade to the tapped hole, press [⏎] to repeat the Mate constraint. Select one of the top arcs of the slot and the tapped hole, as shown in Figure 9.80, to align the centerlines. Press [⏎] to accept the default distance of zero.

Figure 9.80

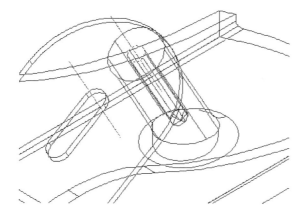

70. The last step is to position the blade handle. We will not use the Insert constraint because the planes and the centerlines are on different parts. Instead, we will mate the two faces and centerlines in two steps. Press [⏎] to repeat the Mate constraint and select the bottom of the handle and the top of the blade as shown in Figure 9.81. Press [⏎] to accept the default offset distance of zero.

Figure 9.81

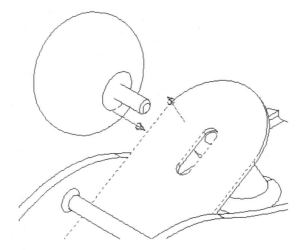

71. To better select the geometry, move the parts so that they are not touching one another. Use the Mate constraint to align the two centerlines of the handle and the tapped hole. Select the center of the handle and the tapped hole as shown in Figure 9.82. Press [⤶] to accept the default offset distance of zero. When complete, the finished assembly should look like Figure 9.83.

Figure 9.82

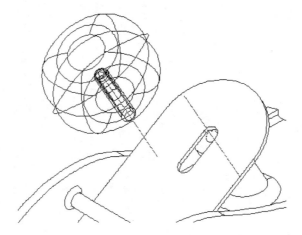

72. Save the file as \Md2book\Drawings\Chapter9\PlaneAssembly.dwg.

Practice Exercises

Figure 9.83

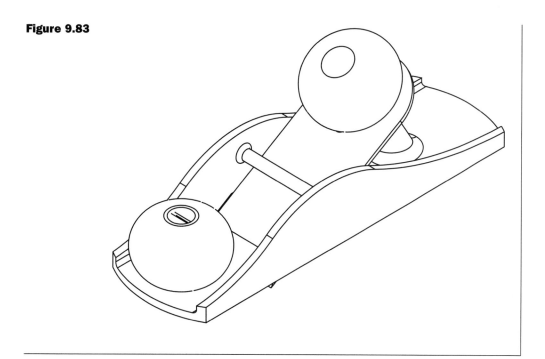

chapter 10

Introduction to NURBS Surfaces

This chapter introduces you to NURBS surfaces and the surface module of Mechanical Desktop through ten tutorials. Eight tutorials will take you through the most common methods of generating NURBS surfaces. The ninth tutorial will show you how Mechanical Desktop uses a NURBS surface to remove material from a parametric solid. Finally, the tenth tutorial will show you how to create drawing views from a surface model.

After completing this chapter, you will be able to:

- Understand the basic principles of surface modeling.
- Understand the major steps required to generate NURBS surfaces.
- Create drawing views from a surface model.

Up to now, the book has covered 3D parametric solids, objects that contain dimensional information throughout the part. A surface model is a representation of the exterior or interior surface(s) of the model, with no dimensional information about what lies inside or outside that particular surface. For example, if you constructed a model using wire and then wrapped the wire with plastic wrap, you would have a realistic surface model. The most common question: why create a surface model when I can create a 3D parametric solid? Not all models can be created using solid modeling. For instance, if you had a model that needed to blend from a circle into a square, this could not be done in MDT 2.0 and would need to be created as a surface model.

A NURBS (non-uniform rational B-spline) surface is based on NURBS mathematics that will accurately define most common free-form surfaces. When you create a NURBS surface, there are three levels of surface continuity that can be created: C0, C1 and C2.

A C0 surface is a surface that has two edges that come together in a discontinuous manner. A good example is a roof or an "L" shaped surface with no bend radius as the "L" joint. Anytime there is a bend in a surface with no blend or radius, a surface is said to be "C0", or discontinuous. Mechanical Desktop does not allow you to generate a C0 surface and it will break the surfaces down into individual surfaces that are C1 or C2.

A C1 surface describes a surface that has a change in continuity, yet is not discontinuous. A good example is a surface that is curved, then flattens out to become planar. Because the surface's continuity changes along a tangency line, the surface is a C1 surface.

A C2 surface is continuous throughout. If you drew a spline and extruded a surface from that spline, you would have a C2 surface. Do not confuse a NURBS surface with the surfaces that can be created in AutoCAD. The surfaces created in AutoCAD are faceted surfaces, made up of triangles and rectangles. Those surfaces do not contain the accuracy and smoothness of a NURBS surface. If you blended a polyface mesh surface from a circle to a square and then looked closely at the edges of the surface, you would find that the edges are actually made up of many flat edges, since they are made up of triangles or rectangles.

Figure 10.1 shows a NURBS and a faceted surface model of a circle blending to a square. The NURBS surface model is smooth all the way around and up the part. In the NURBS model, you see lines representing the surface; these are referred to as flow lines. The number of flow lines on a model does not change the accuracy of the model. Flow lines are merely used for visualizing surfaces. The number of flow lines can be adjusted higher or lower depending on the number of flow lines that you want to see. The reason that the number of flow lines can be adjusted is that a NURBS surface is mathematically calculated and every point on that surface is understood by the model, even if the point is not on a flow line.

The faceted surface model is made up of a series of rectangles (there could also be triangles, depending on the geometry). With a faceted surface, what you see is what you get. Before creating the mesh model, you can increase or decrease the SURFTAB1 and SURFTAB2 variables. This will increase or decrease the number of facets, making the model less or more accurate. The surface information between the rectangles and triangles is assumed to lie in that particular plane. The number of rectangles and triangles cannot be adjusted once the model has been created.

A surface model (NURBS or faceted) is not a parametric model and is not driven by dimensions like parametric solids. Every good surface model starts with a good wireframe model. Wireframe refers to lines, arcs, circles, etc. created in 3D space. For this chapter, you will find the wireframe models for each tutorial are already completed. If you decide to work with NURBS surfaces, you will want to learn more about creating wireframe models. The intent of this chapter is to introduce you to NURBS surfaces and the different types of geometry that can be generated. If, after going through this chapter, you want to learn more about NURBS surfaces, you can refer to the online help.

You will find that generating a surface model follows many of the same steps that you did in generating a parametric solid. First create a wireframe model, similar to creating a sketch. Then generate a surface over the wireframe by extruding, revolving or sweeping the wireframe geometry. Last, create drawing views and annotate the views. When the

Introduction to NURBS Surfaces

Figure 10.1

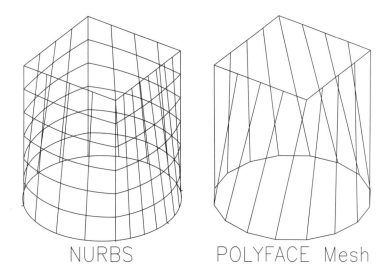

views are generated, no dimensions will appear. However, you can create reference dimensions once the views are created. There is no surface toolbar with the express toolbars. However, you can go to the Surface pull-down menu and select Launch Toolbar and a Surface Modeling toolbar will appear.

Tutorial 10.1—Extruding

The most basic NURBS surface is an extruded surface. Just as a profile is extruded along the Z axis, a cross-section may also be extruded in the Z direction, or along any unidirectional wireframe geometry, to create a NURBS surface.

1. Open the file \Md2book\Drawings\Chapter10\Ex10-1.dwg.
2. Issue the Extruded Surface command (AMEXTRUDESF).
3. At the prompt:

    Select wires:

    select the polyline and press [↵] to tell the system you only want to extrude this geometry.

4. For the direction to extrude the surface, press Z and [↵].
5. Press [↵] to accept the extrusion distance of "1".

Figure 10.2

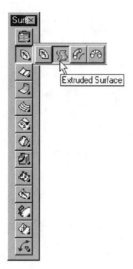

6. Press [⏎ Enter] to accept the default extrusion direction.

7. Press [⏎ Enter] to accept the Taper angle of 0. When complete, your drawing should look like Figure 10.3.

8. Save the file.

Figure 10.3

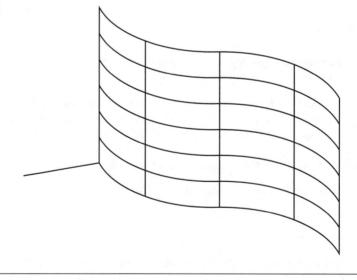

Introduction to NURBS Surfaces

Tutorial 10.2—Editing a Surface

In the previous tutorial, you created an extruded surface. You may have noticed a small line pointing outward from the surface. This is referred to as a surface normal. Every NURBS surface has a surface normal, indicating the position or positive side of the surface. The surface normal is important when NURBS surfaces are milled, since only one side of a surface can be milled (the positive side). The size of the surface normal is purely for viewing purposes. It has no effect on the surface itself and can be adjusted. Figure 10.3 shows the surface normal in the lower left corner of the surface. In this tutorial, you change the size and direction of the surface normal, change the number of flow lines and physically change the shape of the surface using grips.

Figure 10.4

1. Open the file \Md2book\Drawings\Chapter10\Ex10-2.dwg.

2. Issue the Surface Display command (AMDISPSF) and select the surface.

3. In the dialog box, change the surface normal length to .06, change the number of U-Lines to 10 and the number of V-Lines to 6, and then select OK to exit the command.

4. To flip the surface normal, issue the Flip Surface Normal command (AMEDITSF). Select the surface and then select the Normal direction radial button in the dialog box. Press [⏎] and the surface normal will flip to the other side of the surface.

5. Repeat the Flip Surface Normal command, select the surface and press D and [⏎] to flip the normal back. This works only if CMDDIA is set to 0.

Figure 10.5

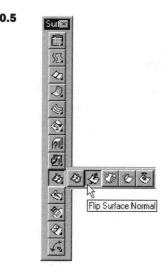

6. If grips are not enabled, turn them on by typing GRIPS at the command line. Press [⏎] and change the value to "1".

7. Select the surface and the grip markers (referred to as control points) will appear on the screen.

8. Select a grip near the middle of the top edge of the surface.

9. To move that control point back and change the appearance of the surface, type in "@.25<90" and press [⏎]. When complete, your drawing should look like Figure 10.6.

Figure 10.6

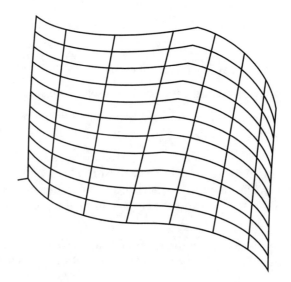

Introduction to NURBS Surfaces

10. Use the dynamic rotation option to view the new surface.
11. Grip edit other points to see how they change the surface.
12. Save the file.

Tutorial 10.3—Creating a Swept Surface

With NURBS surfaces, an object can be swept along a 3D path (the path doesn't need to be 2D or coplanar). A swept surface requires at least one cross-section (but you may have as many as you want), and one or two rails that the cross-section(s) will follow.

In this example we will create a spring. The lisp routine used to create the rail is named spiral.lsp and is included on the CD in the subdirectory Lisp.

Figure 10.7

1. Open the file \Md2book\Drawings\Chapter10\Ex10-3.dwg.
2. Issue the Swept Surface command (AMSWEEPSF).

   For the cross-section, select the circle and press ⏎ to tell AutoCAD that this is the only cross-section to be swept.

3. For the rail, select the spiral and press ⏎.

4. A dialog box allows you to change how the surface will be created. Select OK to accept the defaults. When complete, your drawing should look like Figure 10.8.

5. Save the file.

Figure 10.8

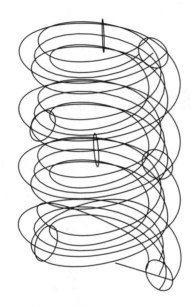

Tutorial 10.4—Blending a Swept Surface

In this example you will blend a surface between two cross-sections, a circle and a square, and have it blend following an arc.

1. Open the file \Md2book\Drawings\Chapter10\Ex10-4.dwg.

2. Issue the Swept Surface command (AMSWEEPSF), (see Figure 10.7).

3. For the cross-section, select the circle and then the square. Press ↵ to stop selecting cross-sections.

4. For the rail, select the arc and press ↵.

5. A dialog box allows you to change how the surface will be created. Select OK to accept the defaults. Another dialog box will appear stating that 5 surfaces will be created; select OK. When complete, your drawing should look like Figure 10.9.

Figure 10.9

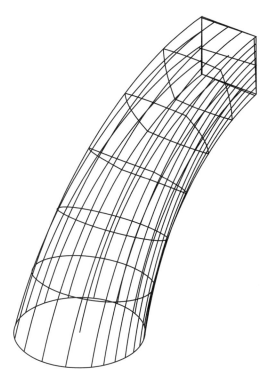

6. To join the surfaces, use the Join Surfaces command (AMJOINSF). Select the five surfaces. The surfaces will need to be selected one at a time; you cannot use a window or crossing method. Press [⏎] when done and your drawing should look like Figure 10.11.

Figure 10.10

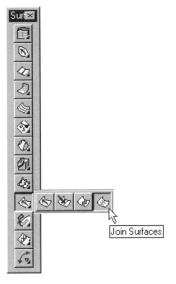

Figure 10.11

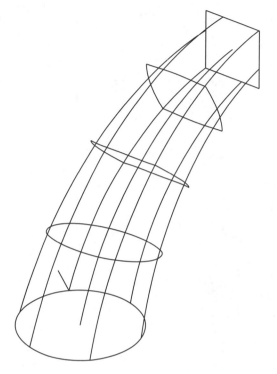

 Tutorial 10.5—Creating a Loft U Surface

A loft U surface is created when multiple cross-sections are facing in the same direction; they do not cross or form a grid. A minimum of two cross-sections is required. In this example, we will create the outside surface of a blender.

1. Open the file \Md2book\Drawings\Chapter10\Ex10-5.dwg.

2. Issue the Loft U Surface command (AMLOFTU) and select the cross-sections in order, top to bottom or bottom to top. The surface will blend from selected cross-section to selected cross-section. When done selecting all the cross-sections, press [↵].

3. A dialog box will appear; select OK to accept the defaults. Your drawing should look like Figure 10.13.

4. Save the file

Introduction to NURBS Surfaces

Figure 10.13

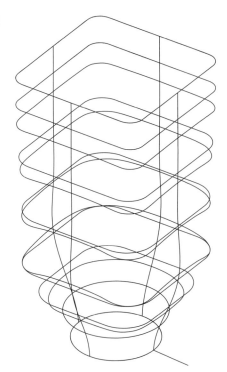

 Tutorial 10.6—Creating a Tubular Surface

To create surfaces that represent tubes or pipes, use the Tubular Surface command (AMTUBE). Before issuing the command, create a 3D polyline that represents the path of the tube. The 3DPOLY command will not allow fillets to be created. When the tube is created, you will have the option to create fillets. Issue the Tubular Surface command, select a 3D polyline and specify the diameter of the tube and the appropriate bend radius, or radii, that you want.

1. Open the file \Md2book\Drawings\Chapter10\Ex10-6.dwg.
2. Issue the Tubular Surface command (AMTUBE) and select the 3D polyline.
3. Type in a Tube diameter of ".25" and press [Enter].
4. Press A for Automatic.
5. Type in a bend radius for all of ".25" and press [Enter]. When complete, your drawing should look like Figure 10.15.
6. Dynamically rotate the model, using the shaded mode.

Figure 10.14

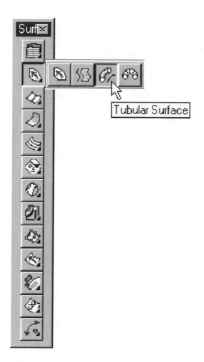

7. Save the file.

Figure 10.15

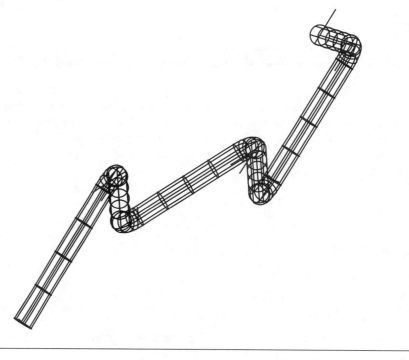

Introduction to NURBS Surfaces

Tutorial 10.7—Creating Primitive Shapes

In this tutorial you will create primitive shapes of a Cone, Cylinder, Sphere and Torus. The Primitive Surface command (AMPRIMSF) will prompt you for locations/dimensions and then the surface model will be built using this information.

Cone

Figure 10.16

1. Start a new drawing.
2. Change to an isometric view (8).
3. Issue the Primitive Surface command (AMPRIMSF) with the Cone option.
4. Select a center point near the left corner of your screen.
5. For the Diameter/Radius prompt, type in ".5" and press [Enter].
6. For the prompt Diameter/Radius of top, press [Enter] to accept the default of "0":
7. Press [Enter] to accept the default for the Height prompt of "1".
8. Press [Enter] to accept the start angle of "0".
9. Press [Enter] to accept the start angle of "0" and also accept the default for Included angle of Full circle. Your cone should look like Figure 10.20.
10. Save the file as \Md2book\Drawings\Chapter10\Ex10-7.dwg.

Figure 10.17

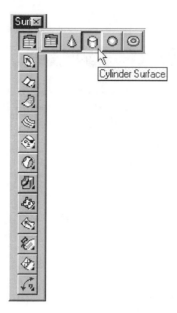

Cylinder

1. In the same drawing as the cone, issue the Primitive Surface command (AMPRIMSF) with the Cylinder option.

2. Select a Base center point to the right of the cone.

3. For the Diameter/<Radius> of base type in ".5" and press [↵].

4. For the Height type in ".5" and press [↵].

5. For the Start angle, press [↵] to accept the default of "0".

6. Press [↵] to accept the Included angle default of Full circle. Your drawing should look like Figure 10.20.

7. Save the file.

Sphere

1. In the same drawing as the cone and cylinder, issue the Primitive Surface command (AMPRIMSF) with the Sphere option.

2. Select a Center point to the right of the cylinder.

3. Type in a Diameter/Radius of the sphere of ".5".

4. Press [↵] to accept the default Start angle of "0".

5. Press [↵] to accept the Included angle default of Full circle. Your drawing should look like Figure 10.20.

6. Save the file.

Introduction to NURBS Surfaces

Figure 10.18

Torus

1. In the same drawing as the cone, cylinder and sphere, issue the Primitive Surface command (AMPRIMSF) with the Torus option.

2. Select a Center point below the cone.

3. Type in a Diameter/Radius of the torus of ".5".

Figure 10.19

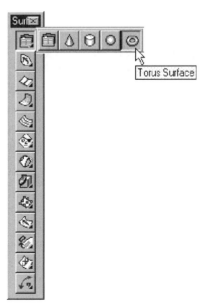

4. Type in a Diameter/Radius of the tube of ".125".

5. Press [↵] to accept the default Start angle of "0".

6. Press [↵] to accept the Included angle default of Full circle. Your drawing should look like Figure 10.20.

7. Save the file.

Figure 10.20

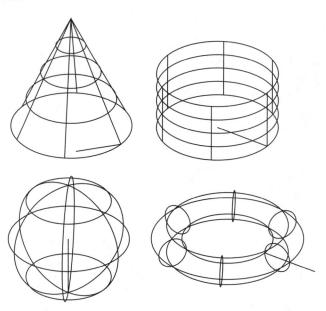

 Tutorial 10.8—Creating a Revolved Surface

To create a revolved surface, you must create a cross-section that will be revolved about an axis. The axis may be a line or selected points.

1. Open the file \Md2book\Drawings\Chapter10\Ex10-8.dwg.

2. Issue the Revolved Surface command (AMREVOLVESF).

3. Select the green polyline for the path curve and then press [↵].

4. For the prompt:

   Axis of revolution:

   press W and [↵] for wire and select the blue vertical line.

5. Press [↵] to accept the start angle of "0".

Introduction to NURBS Surfaces

Figure 10.21

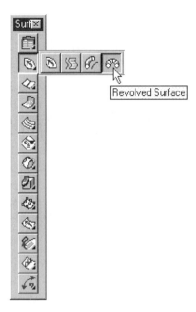

6. Press [⏎] to accept the Included angle default of Full circle. Your drawing should look like Figure 10.22. Note: a dialog box will appear to inform you that two surfaces have been created.

7. Save the file.

Figure 10.22

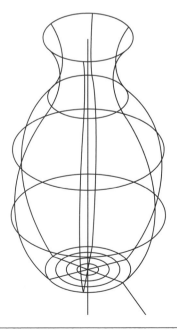

Mechanical Desktop 2.0: Applying Designer and Assembly Modules

Tutorial 10.9—Performing a Surface Cut

In this tutorial, you will perform a surface cut. This will involve using a surface as a cutting face to remove material, or even add material, to a Mechanical Desktop part. (There is also the AMSOLCUT command, which works on AutoCAD native solids.) Before a surface cut is performed, there must be a Mechanical Desktop part and a single surface. After the AMSURFCUT command is issued, there is an option for type. There are 2 types: cuts and protrusions: press T for the "type" option at the command line.

Press P for Protrusion. When a surfcut protrusion is performed, all edges of the surface must be fully contained in the interior of the Mechanical Desktop part. With protrusion, the surface will add material to the part.

Press C for the Cut option. For the Cut option, all of the surface boundaries must extend beyond the faces of the part and the surface will be used as a face to remove material from the part. In this tutorial, a surface model representing a faucet and a Mechanical Desktop part will be used to illustrate how the Surface Cut command works.

Figure 10.23

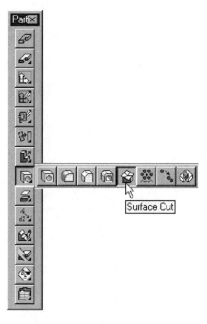

1. Open the file \Md2book\Drawings\Chapter10\Ex10-9.dwg.
2. Issue the Surface Cut command (AMSURFCUT) and select the green surface.

Introduction to NURBS Surfaces

3. An arrow will appear, facing downward. Press F to flip the arrow so that it points upward and press [Enter] to accept this direction. The arrow shows the side to remove. In this case, we will remove the outside material.

4. Use the Dynamic rotation to view the model, selecting the shaded mode. When complete, your drawing should look like Figure 10.24.

5. Save the file.

Figure 10.24

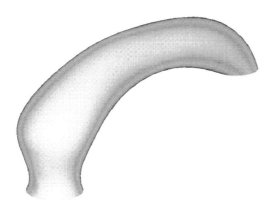

Tutorial 10.10—Creating Drawing Views from a Surface Model

In this tutorial, you will generate drawing views of a blender. The surfaces were placed into a group called Surfaces. When you create drawing views of surfaces, it is best to use groups, because you can add and remove geometry from the group and the drawing views will reflect the geometry that is included in the group. Besides using groups, you will create drawing views in the same manner that you would solids. The drawing views may not be as accurate as views created from a solid model.

1. Open the file \Md2book\Drawings\Chapter10\Ex10-10.dwg.

2. Issue the Create Drawing View command (AMDWGVIEW).

3. Select Group for the Data Set, .25 for the Scale and then select OK.

4. For the Group name to be included, type "Surfaces".

5. For the plane, press U for UCS and then press [Enter].

6. Press [Enter] to accept the default orientation.

Mechanical Desktop 2.0: Applying Designer and Assembly Modules

7. Select a point in the upper left corner of the border.
8. Place in two Ortho views and a Iso view, as shown in Figure 10.25.

Figure 10.25

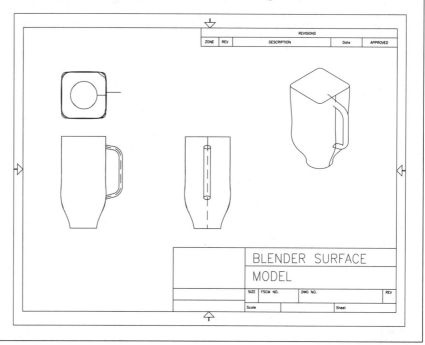

Tip: To increase the performance of the view generation, you can turn off Calculate Hidden Lines before creating drawing views.

The previous ten tutorials should have given you a good overview of the potential for surfaces. If you want to learn more about NURBS surfaces, refer to the online help.

Review Questions

1. A surface model contains information about what lies inside the surface area. T or F?
2. A surface model is not parametric. T or F?
3. What are C0, C1 and C2 surfaces?
4. The direction of a surface normal cannot be changed. T or F?
5. What is the difference between a NURBS and a faceted surface?
6. A swept surface can only have one cross-section. T or F?
7. Once a surface has been created, it cannot be edited. T or F?
8. What are the four types of surface primitives you can generate?
9. Multiple surfaces can be selected for performing a surface cut. T or F?
10. What Data set should you use to generate drawing views from surface models?

glossary

AM2SF Surface. Changes objects into surfaces.

AMABOUT Displays version and copyright information.

AMACTIVATE Assembly. Selects and activates a specific part, assembly, or scene in the Part/Assembly environment.

AMADDCON Part. Adds parametric constraints to the selected profile, path, or cutting line sketch.

AMANGLE Assembly. Creates alignment constraints between two planes to form specified angles.

AMANNOTE Drawing. Creates, adds, moves, or deletes drawing annotations.

AMARRAY Part. Creates rectangular and polar arrays of extrusions, revolutions, sweeps, holes, and combine part features.

AMASSEMBLE Assembly. Solves assembly constraints and updates the part locations in the assembly.

AMASSIGN Assembly. Gives a value to a part attribute or adds a new user-defined attribute to the part definition.

AMASSMPROP Assembly. Generates mass property information about a part.

AMAUDIT Assembly. Lists externally referenced drawing files that need updating.

AMAUGMENT Surface. Creates augmented lines from surface edges, internal trim lines, and display lines.

AMBALLOON Assembly. Creates and places balloons in drawing views of scenes.

AMBLEND Surface. Creates a surface between two, three, or four wires or surfaces.

AMBOM Assembly. Creates and places a bill of materials table.

AMBOMSETUP Assembly. Controls the format for the display of bills of materials and balloons.

AMBREAK Surface. Breaks one surface into two surfaces along a specified direction.

AMBROWSER Provides a visual structure for each part, assembly, scene, or drawing with icons that represent a part, feature, constraint, or attribute as it is added to a model.

AMCATALOG Assembly. Maintains a catalog of all standard parts, both local and attached, currently being used in the assembly model.

AMCENLINE Drawing. Creates a parametric centerline in Drawing mode.

AMCHAMFER Part. Creates a chamfer on the selected edge(s) of the active part.

AMCHECKFIT Surface. Measures the distance between objects.

AMCOMBINE Part. Combines the active base part with another parametric part, the toolbody, within the active assembly.

AMCONSTRAIN Assembly. Creates, modifies, and removes constraints in a 3D copy.

AMCOPYFEAT Part. Copies sketched and hole features or feature sets from within the same part or from one part to another.

AMCOPYSKETCH Part. Selects single or multiple sketches to copy and specifies placement of the sketch center.

AMCORNER Surface. Creates a blended fillet corner surface from three intersecting fillet surfaces.

AMCUTLINE Part. Creates a cutting line sketch for an offset section view.

AMDATUMID Mechanical Drafting. Creates and edits datums represented in frames containing an alphabetic character.

AMDATUMTGT Mechanical Drafting. Creates and edits datum target symbols.

AMDELCON Part. Removes constraints from the selected sketch.

AMDELETE Part. Deletes part definitions, part instances, and scenes from the current assembly model.

AMDELFEAT Drawing. Deletes features from the selected part.

AMDELTRAIL Assembly. Deletes a selected trail.

AMDELTWEAKS Assembly. Deletes all tweaks associated with the selected parts.

AMDELVIEW Drawing. Deletes specified views and all views dependent upon them.

AMDIMALIGN Mechanical Drafting. Aligns linear dimensions to be parallel and angular dimensions to be concentric.

AMDIMBREAK Mechanical Drafting. Creates a gap (break) between two points on an existing associative dimension or extension line.

AMDIMDSP Part. Changes the display mode for dimensions.

AMDIMFORMAT Edits individual dimensions with the dialog box.

AMDIMINSERT Mechanical Drafting. Inserts a new dimension simultaneously.

AMDIMJOIN Mechanical Drafting. Joins two or more linear or angular dimensions.

AMDIRECTION Surface. Displays the direction of objects and reverses their direction.

AMDISPSF Surface. Controls how surfaces are displayed.

AMDIST Surface. Measures the minimum 3D distance between two sets of objects.

AMDWGDIMDSP Drawing. Changes the display of parametric dimensions in Drawing mode.

AMDWGOUT Drawing. Converts Mechanical Desktop data to AutoCAD 2D data.

AMDWGVIEW Drawing. Creates base, orthographic, auxiliary, isometric, detail, section, and broken views.

AMEDGE Surface. Untrims, copies, and extracts edges of surfaces and shows edge nodes of surfaces and faces.

AMEDIT Mechanical Drafting. Edits symbols for welding, surface textures, feature control frames, datum targets, and datum and feature identifiers.

AMEDITAUG Surface. Resizes, rotates, copies, and corrects vectors on augmented lines and creates augmented lines from lines and polylines.

AMEDITBOM Assembly. Shows and manages the active bill of materials and its entries.

AMEDITCONST Assembly. Modifies and deletes constraints in 3D assemblies.

AMEDITFEAT Part. Displays and modifies dimension values for features of the active part.

AMEDITSF Surface. Changes the density of surface grip points, the grip span, and the direction of the normal and truncates surfaces.

AMEDITTRAIL Assembly. Modifies the explosion path start and end offsets in an assembly scene.

AMEDITVIEW Drawing. Modifies the scale, text, associativity, and hidden line display of a drawing view.

AMEXTRUDE Part. Creates an extruded solid feature.

AMEXTRUDESF Surface. Creates a surface by extruding a line, arc, circle, ellipse, spline, or polyline.

AMFCFRAME Mechanical Drafting. Creates a feature control frame.

AMFEATID Mechanical Drafting. Creates a feature identifying symbol.

AMFILLET Part. Creates and displays an example of a specified fillet type.

AMFILLET3D Surface. Creates a smoothly fitted arc of constant radius between wires.

AMFILLETSF Surface. Creates a fillet surface along the intersection of two surfaces.

AMFITSPLINE Surface. Smoothes wire objects by fitting them with a smooth NURBS spline.

AMFIXPT Part. Fixes a point on a sketch in XYZ space.

AMFLOW Surface. Creates flow wires in the U and V directions.

AMFLUSH Assembly. Creates a parallel at a specified offset between two planes.

AMHOLE Part. Creates drilled, counterbored, or countersunk holes that may be termi-

Glossary

nated on planar faces.

AMHOLENOTE Part. Creates custom diameter, depth, and angle information for standard hole notes.

AMIDFIN Reads printed circuit board (PCB) data in the Intermediate Data Format (IDF) and converts it to AutoCAD and Mechanical Desktop objects.

AMINSERT Assembly. Creates a constraint to make two circular edges share the same center point and to make their planes coplanar.

AMINTERFERE Assembly. Checks for interference among parts within an assembly.

AMINTERSF Surface. Intersects two surfaces and creates a 3D polyline at the intersection.

AMJOIN3D Surface. Joins wires to form one 3D polyline, spline, or augmented line.

AMJOINSF Surface. Joins two or more surfaces at their untrimmed base edges into one continuous surface.

AMLENGTHEN Surface. Extends or shortens base surface edges by a percentage or distance.

AMLISTASSM Assembly. Returns information about component instances.

AMLISTDWG Drawing. Lists view information in a text window.

AMLISTPART Part. Displays part, feature, and view information.

AMLOFTU Surface. Creates a surface through a set of wires.

AMLOFTUV Surface. Creates a surface through two sets of wires.

AMMAKEBASE Part. Converts an active part into a static base feature.

AMMATE Assembly. Creates a mate constraint between two parts.

AMMIRROR Assembly. Generates a mirrored copy of a part in the Single Part and Part/Assembly environments.

AMMODDIM Part. Modifies parametric dimensions on sketches and drawings.

AMMODE Drawing. Controls whether Model or Drawing mode is in effect.

AMMODIN Surface. Loads files from previous releases.

AMMODOUT Surface. Formats files for use in AutoSurf/AutoMill Release 6.0.

AMMOVEDIM Drawing. Controls location of dimensions in Drawing Mode.

AMMOVEVIEW Drawing. Moves drawing views in Drawing mode.

AMNEW Assembly. Creates new instances of parts, scenes, and subassemblies.

AMOFFSET3D Surface. Creates copies of 3D polylines that are set off from the original.

AMOFFSETSF Surface. Creates a surface offset from a selected surface.

AMPARDIM Part. Creates parametric dimensions in the Model environment.

AMPARTEDGE Part. Creates a line on a selected edge to use in a new profile.

AMPARTIN Assembly. Places an external part in the current open file.

AMPARTLINE Surface. Creates a 3D polyline on the profile of the surface in the current view.

AMPARTOUT Assembly. Places a part or subassembly in an external file.

AMPARTPROP Part. Lists the mass properties for a selected part.

AMPATH Part. Creates a constrained sketch used as a path for a sweep feature.

AMPATTERNDEF Assembly. Sets hatching pattern attributes for part definitions.

AMPLANE Surface. Creates one or more planar surfaces.

AMPREFS Manages part, assembly, surface, drawing, and Desktop preferences from a single dialog box.

AMPRIMSF Surface. Creates a cone, cylinder, torus, and sphere.

AMPROFILE Part. Solves 2D geometry to create a constrained sketch used as a profile for extrude, revolve, or sweep features.

AMPROJECT Surface. Projects a wire onto a surface.

AMREFDIM Drawing. Creates reference dimensions in Drawing mode.

Glossary

AMREFINE3D Surface. Changes the point density of lines and 3D polylines.

AMREFINESF Surface. Redefines the selected surface with a less complex approximation.

AMREFRESH Assembly. Updates external part or subassembly definitions.

AMRENAME Assembly. Renames an active part definition, part instance, or scene.

AMREORDFEAT Part. Changes the order of a feature in the part creation history.

AMREPLACE Assembly. Changes definitions for part or subassembly instances.

AMREPLAY Part. Redisplays, in sequence, the steps used to create features and sketches of the selected part.

AMREVOLVE Part. Creates a revolved solid feature from the selected profile.

AMREVOLVESF Surface. Creates a surface by rotating a path wire around a selected axis.

AMRULE Surface. Creates a straight element surface between two wires.

AMSCALE Surface. Increases or decreases individual or all surface axes by a specific scale factor.

AMSECTION Surface. Creates section cuts through one or more surfaces.

AMSHELL Part. Creates a shell with an assigned wall thickness or multiple wall thicknesses.

AMSHOWACT Part. Highlights the active part or sketch plane.

AMSHOWCON Part. Displays the constraint symbols on the selected sketch.

AMSHOWINST Assembly. Highlights selected part instances in the Desktop Browser.

AMSKPLN Part. Sets the sketch plane location and XY axis orientation.

AMSOLCUT Surface. Cuts an AutoCAD solid model with a Mechanical Desktop surface.

AMSTYLEI Mechanical Drafting. Imports existing dimension styles.

AMSUPPRESS Assembly. Suppresses the display of instances in a scene.

AMSURFCUT Part. Cuts free-form surface shapes on solid models.

AMSURFPROP Surface. Calculates the mass properties of surfaces.

AMSURFSYM Mechanical Drafting. Adds surface texture finish symbols to drawings.

AMSWEEP Part. Creates a solid feature from a profile moved along a path.

AMSWEEPSF Surface. Creates a surface by sweeping cross sections along one or two rails.

AMSYMSTD Mechanical Drafting. Edits and defines drafting standards for symbols.

AMTEMPLATE Part. Creates, renames, and edits templates for hole notes.

AMTRAIL Assembly. Adds an explosion path to an assembly scene.

AMTUBE Surface. Creates a tubular surface around a selected wire that becomes the axis of the tube.

AMTWEAK Assembly. Adjusts the position of a part or subassembly in an exploded assembly scene.

AMUNSPLINE Surface. Changes splines to fit points or to polylines.

AMUPDATE Part. Regenerates the active part and drawing with new values, dimensions, and constraints.

AMUPDATESCENE Assembly. Revises the current scene to reflect editing.

AMUPDATEDWGVIEW Drawing. Updates drawing views.

AMVARS Part. Creates and manages active part and global design variables and provides table-driven parts by linking to Microsoft Excel spreadsheets.

AMVIEW Controls viewing in model space.

AMVISIBLE Controls the visibility of parts, assemblies, scenes, drawings, and geometric objects.

AMVRMLOUT Converts selected objects to VRML (Virtual Reality Modeling Language) to be viewable on a web page.

Glossary

AMWELDSYM Mechanical Drafting. Creates a welding symbol.

AMWHEREUSED Assembly. Shows where a part or subassembly definition is used.

AMWORKAXIS Part. Creates a work axis at the centerline of a cylindrical, conical, or toroidal surface.

AMWORKPLN Part. Creates a construction plane on the selected part.

AMWORKPT Part. Creates work points on the active sketch plane.

AMXFACTOR Assembly. Controls the explosion factor for scenes.

AV3DVARS Changes the user-configurable 3D display options.

AVEDGES Displays edges on shaded graphics.

AVHIDE Sets the 3D display mode to hidden line.

AVMAT Displays the 3D Graphics Material Values dialog box.

AVPAN Moves the 3D display in the current viewport along the view plane.

AVRENDER Sets the 3D display for rendering.

AVROTATE Performs 3D view rotation of a model about its center of geometry.

AVROTATEPT Rotates a model about a user-specified point.

AVSELECT Activates the cursor for selecting the items to be included in the 3D display.

AVWIREFRAME Sets the 3D display mode to wireframe.

AVZOOM Increases or decreases the 3D display magnification of the model.

DDVIEW AutoCAD command for saving views.

DVIEW AutoCAD command that takes you out of render mode with additional view options.

MCAD1 Displays the screen as one viewport.

MCAD2 Divides the screen into two viewports.

MCAD3 Divides the screen into three viewports.

MCAD4 Divides the screen into four viewports.

PAN AutoCAD command that pans in wireframe mode and works in realtime.

VIEW Saves and restores the first, second, and third views of a drawing.

VPOINT Sets the viewing direction for a three-dimensional visualization of the drawing.

ZOOM AutoCAD command that zooms in wireframe mode and works in realtime.

Glossary ©1997 by Autodesk, Inc.

index

A

Accept option, 60

Active Part Mass Properties, 207

Active part variable, 153-54, 156-57. (*See also* Design variables)

Active part, 182-83, 259, 308

Add Dimension, 13-16

Add Tweaks, 248-50, 253

Adding constraints, 11-12

Adding parts, 179-80

Align Dimension, 282-83

Align horizontal, 269

Align vertical, 269

AM_BL. (*See* Layer names)

AM_BM. (*See* Layer names)

AM_HID. (*See* Layer names)

AM_PARDIM. (*See* Layer names)

AM_REFDIM. (*See* Layer names)

AM_TR. (*See* Layer names)

AM_VIEWS. (*See* Layer names)

AM_VIS. (*See* Layer names)

AM_WORK. (*See* Layer names)

AMANGLE. (*See* Angle constraint)

AMARRAY. (*See* Arrays)

AMASSMPROP. (*See* Assembly Queries)

AMBALLOON. (*See* Balloons)

AMBOMSETUP. (*See* Create a BOM Table)

AMBROWSER. (*See* Desktop Browser)

AMCATALOG. (*See* Part Catalog)

AMCENLINE. (*See* Centerline)

AMCHAMFER. (*See* Chamfers)

AMCOMBINE. (*See* Combine)

AMCONSTRAIN. (*See* Constraints)

AMCOPYFEAT. (*See* Feature, copying)

AMCOPYSKETCH. (*See* Copy Sketch)

AMDELCON. (*See* Delete Constraint)

AMDELFEAT. (*See* Delete Feature)

AMDELTRAIL. (*See* Delete Trail)

AMDELTWEAKS. (*See* Delete Tweaks)

AMDELVIEW. (*See* Delete Drawing View)

AMDIMALIGN. (*See* Align Dimension)

AMDIMBREAK. (*See* Break Dimension)

AMDIMDSP. (*See* Dimension Display)

AMDIMINSERT. (*See* Insert Dimension)

AMDIMJOIN. (*See* Join Dimensions)

AMDISPSF. (*See* Surface Display)

AMDRGVIEW. (*See* Create Drawing View)

AMEDITCONST. (*See* Edit Constraints)

AMEDITFEAT. (*See* Edit Feature)

AMEDITFORMAT. (*See* Edit Format)

AMEDITSF. (*See* Flip Surface Normal)

AMEDITTRAIL. (*See* Edit Trail)

AMEXTRUDE. (*See* Extrude)

AMEXTRUDESF. (*See* Extruded Surface)

AMFILLET. (*See* Fillets)

AMFIXPT. (*See* Fix Point)

AMFLUSH. (*See* Flush constraint)

AMHOLE. (*See* Hole)

AMHOLENOTE. (*See* Hole, notes)

AMINSERT. (*See* Insert constraint)

AMINTERFERE. (*See* Interference checking)

AMJOINSF. (*See* Join Surfaces)

AMLOFTU. (*See* Loft U Surface)

AMMANNOTE. (*See* Annotation)

AMMATE. (*See* Mate constraint)

AMMIRROR. (*See* Mirroring a part)

AMMODDIM. (*See* Change Dimension)

AMMOVEVIEW. (*See* Move Drawing View)

AMNEW. (*See* New Part)

AMPARDIM. (*See* Add Dimension)

AMPARTPROP. (*See* Active Part Mass Properties)

AMPATH. (*See* Path)

AMPREFS. (*See* Edit Preferences)

AMPRIMSF. (*See* Primitive Surface)

AMPROFILE. (*See* Profile a Sketch)

AMREFDIM. (*See* Dimension, reference)

Index

AMREORDFEAT. (*See* Reorder Feature)

AMREPLAY. (*See* Feature Replay)

AMREVOLVE. (*See* Revolve)

AMREVOLVESF. (*See* Revolved Surface)

AMSCALE. (*See* Scaling a part)

AMSHELL. (*See* Shelling)

AMSHOWCON. (*See* Show Constraints)

AMSKPLN. (*See* Create Sketch Plan)

AMSOLCUT, 354

AMSURFCUT. (*See* Surface Cut)

AMSWEEP. (*See* Sweep)

AMSWEEPSF. (*See* Swept Surface)

AMTRAIL. (*See* Create Trail)

AMTUBE. (*See* Tubular Surface)

AMTWEAK. (*See* Add Tweaks)

AMUPDATE. (*See* Update Part)

AMVARS. (*See* Design variable)

AMVIEW. (*See* Desktop View)

AMVISIBLE. (*See* Part Visibility)

AMWORKAXIS. (*See* Work axis)

AMWORKPLN. (*See* Work Plane)

AMWORKPT. (*See* Work Points)

Angle constraint, 224, 227, 234-36

Angular dimension, 13, 15

Angular Tolerance, 2 (*See also* Part tab options)

Annotation, 285-86

Apply Constraint Rules, 1-2, 7 (*See also* Part tab options)

Apply to Linetype, 3 (*See also* Part tab options)

Arc, dimensioning, 13

Arrays, 135-42

   creating, 137-42

   editing, 136

   polar, 136, 309

   rectangular, 136

Assemblies

   constraints, 221-45

      types, 223-24

   creating, 213-42

      bottom up approach, 218-21

      top down approach, 214-18

Assembly Queries, 208-10

Assume Rough Sketch, 2, 5 (*See also* Part tab options)

Aux view, 259

B

Balloons, 290-292 (*See also* Bill of materials)

Base view, 258, 266

Bill of materials, 288-94

   balloons, 290-92

   table definition section, 288-90

Break Dimension, 282, 284

Broken view, 259

Browser Assembly tab, 101-8

   copying, 104-6

   deleting, 104-6

   editing features, 101-2

   renaming, 102-4

   reordering, 106-8

C

C0 surface, 337

C1 surface, 338

C2 surface, 338

Calculate hidden lines, 260, 268

Center marks, 285, 287

Centerline, 285-87

Chamfers, 117-23, 322

   equal distance, 117-19

   two distances, 119-22

   x angle, 122-23

Change Dimension, 15-16, 148, 170, 273

Charted drawings. (*See* Parametric part)

Circle, dimensioning, 13

Close edge, 163-66

Collinear constraint, 7, 320

Combine, 47, 200-204, 246, 301

Compress, 3

Concentric constraint, 7

Constraint, Part

   adding, 11-12

   deleting, 12

   Size, 3 (*See also* Part tab options)

Constraints, 223-24, 228-41

Construction geometry, 169-72

Converting drawings, 172-74

Copy Sketch, 166-69

Create a BOM Table, 288-94

Create Drawing View, 257-67, 355

   data set, 259

Index

hidden lines, 260

scale, 260-61

section, 259-60

type, 258-59

Create Scene, 247-49

Create Sketch Plan, 59-62, 64

Create Trail, 250-54

Cut, 84-93, 165. (*See also* Sketch features)

D

Data set, drawing view, 259

Delete Constraint, 12, 50

Delete Drawing View, 270-73

Delete Feature, 65, 78

Delete Trail, 252

Delete Tweaks, 249-50

Design variable, 153-63, 156, 158-59

Desktop Browser, 101-8

Desktop Preferences tabs

Assemblies, 1

Desktop, 1

Drawing, 1

Part, 1-3

Surfaces, 1

Desktop View, 85

Desktop View toolbar, 24-25

Detail view, 259

Dimension Display, 147-53

modes, 147-48

Dimension

editing, 273

hiding, 274-75

move, 275-78

reattach, 275-78

reference, 278-79

Dimensioning, 13-19

Display tangencies of hidden lines, 260, 269

Drawing view

editing, 268-69

types, creating, 258-59

Drawing Visibility, 274

DVIEW, 25

Dynamic rotation, 24-25, 26

E

Edge properties, 269

Edit Constraints, 242-45

Edit Feature, 46-47, 49, 65, 87, 108, 123, 200-201

Edit Format, 281-82

Edit Preferences, 102

Edit Trail, 251, 253

Equations. (*See* Dimension Display)

Excel Spreadsheet, 160-63

Excluded faces, 189-90. (*See also* Shelling)

Externalizing, 220

Extrude, 27, 30-31, 48-49, 168, 298, 303, 305

   with cut: blind, 89-90

   with cut: through, 87

   with intersect: through, 99

   with join: blind, 93-94

   with join: to plane, 95-97

Extruded Surface, 339-40

Extruding the profile, 27-32, 84-85

Extrusion Feature dialog box

   Operation, 27-28, 30, 31

   Size, 29, 30, 31

   Termination, 28, 30, 31

F

Faces. (*See* Thickness)

Feature, 83

   copying, 194-97

Feature editing, 45-51

Feature Replay, 204-7, 297

Fillets, 108-17, 299

   constant, 109-10

   cubic, 113-15

   fixed width, 115-17

   linear, 111-13

Fix Point, 6

Flip Surface Normal, 341-42

Flush constraint, 224-25, 231-32

Full section view, 259

G

Global variable, 154-56, 158-59. (*See also* Design variables)

Group, data set, 259

H

Half section view, 259

Hatch pattern section view, 260

Hidden lines, creating drawing views, 260, 267

Hide hidden lines, 260, 268

Hole, 123-35, 328

   C'Bore/Sunk Size, 125

   Drill Size, 125

   notes, 279-81

   Operation, 124

   Placement, 125,

   Tapped, 126

   Termination, 124

Horizontal constraint, 7

I

Independent array instance, 46

Insert constraint, 224, 226, 233-36, 331

Insert Dimension, 282, 284

Instance, 180-82

Interference checking, 245-47

Intersect, 84, 99-100. (*See also* Sketch features)

Iso, 259, 264

J

Join, 84, 93-99, 164, 298, 301. (*See also* Sketch features)

Join constraint, 7

Join Dimensions, 282-83

Join Surfaces, 345-46

L

Layer names, 261

Linear dimension, 15

Linetype of hidden lines, 260, 268

Loft U Surface, 346-47

M

Mass properties, 207-10

Mate constraints, 224-25, 229-31, 237-42, 314, 329-34

Mirroring a part, 183-86

Model Views, 23

Move with parent, 269

Moving drawing views, 269-70

N

Naming Prefixes, 3. (*See also* Part tab options)

New Part, 180, 214

Non-uniform rational B-spline. (*See* NURBS)

Number keys, 24

NURBS surfaces, 337-56

O

Offset section view, 259

Ordinate dimension, 15

Ortho view, 259, 262, 356

Over shoot. (*See* Create Trail)

P

Parallel constraint, 7

Parametric Boolean, 194

Parametric parts, converting to, 172-74

Parametric solid, 1

Part Catalog, 180, 214, 218, 220

Part Modeling toolbar, 5-7, 11

Part tab options, 1-3, 101

Part Visibility, 79-81

Path, 38, 44, 310

Perpendicular constraint, 7

Placement, 269

Primitive Surface, 349-52

Profile a Sketch, 5

Profile, extrude, 27-32

Profiling. (*See* Profile a Sketch)

Project constraint, 7

R

Radial dimension, 15

Radius constraint, 7

Relative to parent, 269

Renumber, 292

Reorder Feature, 197-200

Replaying, 204-7

Revolution Feature dialog box

  Operation, 32-33, 35, 36

  Size, 34, 35, 36

  Termination, 33, 35, 36

Revolve

  a profile, 32-37

  with cut: full, 90-92

with intersect: full, 100

with join: from to, 97-99

Revolved Surface, 352-53

Rotate option, 60, 62, 64

S

Saved File Format, 3. (*See also* Part tab options)

Scale, drawing view, 260-61, 269

Scaling a part, 186-88

Scene, 259

Section symbol, 260

Section views, creating drawing, 259-60

Select, data set, 259

Select feature option, 47-48, 49

Set. (*See* Thickness)

Shelling, 188-93, 299, 305

Show Constraints, 11, 12

Sketch features, 84-100

  operations, 85

  options, 86

  termination, 85

Sketch option, 47

Sketch plane, 59-64

Sketching tips, 4-5

Surface Cut, 354-55

Surface Display, 341

Surface models

  faceted, 338

  NURBS, 337-56

Surface normal, 341-43

SurfCut option, 47

Sweep, 40-42, 44, 311

  Body type, 41-42

  Operation, 40-41

  Size, 42

  Termination, 41

  with cut: path only, 92-93

Sweeping a profile, 38-45, 84-85

Swept surface, 343-44

Symbols, 286-87

T

Table-driven variable, 160-63

Tabulated drawings. (*See* Parametric part)

Tangent constraint, 7, 312

Thickness. (*See also* Shelling)

   default, 188-89

   multiple overrides, 190

Tolerance/Pickbox Size, 2. (*See also* Part tab options)

Toolbody option, 47

Tubular Surface, 347-48

U

UCS. (*See* User-defined coordinate system)

UCSICON, 26

Under shoot. (*See* Create Trail)

Update Part, 46-47, 49, 101, 273

User-defined coordinate system, 26, 39, 57, 59-60, 62, 66

V

Vertical constraint, 7

View properties, 269

Viewing a model, 23-26

Viewports, 24

W

Wireframe model, 338-39

Work axis, 57-59, 303, 308

Work plane, 65-77

  On Edge/Axis, 66-68

  On Edge/Axis – Planar, 68-69, 308

  On Vertex, 75-77

  Planar Parallel – Offset, 74-75

  Planar Parallel – On Vertex, 73-74

  Tangent – On Edge/Axis, 72-73

  Tangent – Planar Parallel, 69-72

Work plane, 39, 43, 45

Work points, 78-79

World Coordinate System, 26

X

Xvalue constraint, 7, 302, 322

XY orientation, 26

Y

Yvalue constraint, 7

Z

Z-Flip option, 60, 62

License Agreement for Autodesk Press
an International Thomson Publishing company
Educational Software/Data

You the customer, and Autodesk Press incur certain benefits, rights, and obligations to each other when you open this package and use the software/data it contains. BE SURE YOU READ THE LICENSE AGREEMENT CAREFULLY, SINCE BY USING THE SOFTWARE/DATA YOU INDICATE YOU HAVE READ, UNDERSTOOD, AND ACCEPTED THE TERMS OF THIS AGREEMENT.
Your rights:
1. You enjoy a non-exclusive license to use the enclosed software/data on a single microcomputer that is not part of a network or multi-machine system in consideration for payment of the required license fee, (which may be included in the purchase price of an accompanying print component), or receipt of this software/data, and your acceptance of the terms and conditions of this agreement.
2. You own the media on which the software/data is recorded, but you acknowledge that you do not own the software/data recorded on them. You also acknowledge that the software/data is furnished "as is," and contains copyrighted and/or proprietary and confidential information of Autodesk Press or its licensors.
3. If you do not accept the terms of this license agreement you may return the media within 30 days. However, you may not use the software during this period.

There are limitations on your rights:
1. You may not copy or print the software/data for any reason whatsoever, except to install it on a hard drive on a single microcomputer and to make one archival copy, unless copying or printing is expressly permitted in writing or statements recorded on the diskette(s).
2. You may not revise, translate, convert, disassemble or otherwise reverse engineer the software/data except that you may add to or rearrange any data recorded on the media as part of the normal use of the software/data.
3. You may not sell, license, lease, rent, loan, or otherwise distribute or network the software/data except that you may give the software/data to a student or and instructor for use at school or, temporarily at home.

Should you fail to abide by the Copyright Law of the United States as it applies to this software/data your license to use it will become invalid. You agree to erase or otherwise destroy the software/data immediately after receiving note of Autodesk Press' termination of this agreement for violation of its provisions.

Autodesk Press gives you a LIMITED WARRANTY covering the enclosed software/data. The LIMITED WARRANTY can be found in this product and/or the instructor's manual that accompanies it.

This license is the entire agreement between you and Autodesk Press interpreted and enforced under New York law.

Limited Warranty

Autodesk Press warrants to the original licensee/purchaser of this copy of microcomputer software/data and the media on which it is recorded that the media will be free from defects in material and workmanship for ninety (90) days from the date of original purchase. All implied warranties are limited in duration to this ninety (90) day period. THEREAFTER, ANY IMPLIED WARRANTIES, INCLUDING IMPLIED WARRANTIES OF MERCHANTABILITY AND FITNESS FOR A PARTICULAR PURPOSE ARE EXCLUDED. THIS WARRANTY IS IN LIEU OF ALL OTHER WARRANTIES, WHETHER ORAL OR WRITTEN, EXPRESSED OR IMPLIED.

If you believe the media is defective, please return it during the ninety day period to the address shown below. A defective diskette will be replaced without charge provided that it has not been subjected to misuse or damage.

This warranty does not extend to the software or information recorded on the media. The software and information are provided "AS IS." Any statements made about the utility of the software or information are not to be considered as express or implied warranties. Delmar will not be liable for incidental or consequential damages of any kind incurred by you, the consumer, or any other user.

Some states do not allow the exclusion or limitation of incidental or consequential damages, or limitations on the duration of implied warranties, so the above limitation or exclusion may not apply to you. This warranty gives you specific legal rights, and you may also have other rights which vary from state to state. Address all correspondence to:

Autodesk Press
3 Columbia Circle
P. O. Box 15015
Albany, NY 12212-5015